SpringerBriefs in Electrical and Computer Engineering

Computational Electromagnetics

Series editor

Rakesh Mohan Jha, Bangalore, India

More information about this series at http://www.springer.com/series/13885

Balamati Choudhury · Arya Menon
Rakesh Mohan Jha

Active Terahertz Metamaterial for Biomedical Applications

 Springer

Balamati Choudhury
Centre for Electromagnetics
CSIR-National Aerospace Laboratories
Bangalore, Karnataka
India

Rakesh Mohan Jha
Centre for Electromagnetics
CSIR-National Aerospace Laboratories
Bangalore, Karnataka
India

Arya Menon
Centre for Electromagnetics
CSIR-National Aerospace Laboratories
Bangalore, Karnataka
India

ISSN 2191-8112   ISSN 2191-8120   (electronic)
SpringerBriefs in Electrical and Computer Engineering
ISSN 2365-6239   ISSN 2365-6247   (electronic)
SpringerBriefs in Computational Electromagnetics
ISBN 978-981-287-792-5   ISBN 978-981-287-793-2   (eBook)
DOI 10.1007/978-981-287-793-2

Library of Congress Control Number: 2015947809

Springer Singapore Heidelberg New York Dordrecht London

Printed on acid-free paper

Springer Science+Business Media Singapore Pte Ltd. is part of Springer Science+Business Media
(www.springer.com)

To Professor Satya N. Atluri

In Memory of Dr. Rakesh Mohan Jha
Great scientist, mentor, and excellent
human being

Dr. Rakesh Mohan Jha was a brilliant contributor to science, a wonderful human being, and a great mentor and friend to all of us associated with this book. With a heavy heart we mourn his sudden and untimely demise and dedicate this book to his memory.

Preface

Terahertz spectroscopy is gaining momentum as a tool for imaging in the field of biomedical engineering. This increase in popularity is due to the non-invasive, non-ionizing nature of terahertz radiation coupled with its propagation characteristics in water, which allows the operator to obtain high-contrast images of skin cancers, burns, etc., without detrimental effects. In order to tap this huge potential, researchers are aiming to build highly efficient biomedical imaging systems by introducing radiation (THz) absorbers into biomedical detectors. The biggest challenge faced in the fulfilment of this objective is the lack of naturally occurring dielectrics, which is combated with the use of artificially engineered resonant materials called metamaterials.

Due to sharp resonance, metamaterial absorbers show narrow band absorption. Therefore, in order to use the same detector for imaging and characterization of a variety of samples at different frequencies, a technique to tune the resonant frequency of the metamaterial absorber is introduced. In this technical brief, the design of an active, metamaterial absorber for biomedical imaging in the terahertz range of frequencies is explored. The design is optimized for near unity absorption using a particle swarm optimization (PSO)-based computational engine.

The design of a terahertz absorber is a three-step process involving integration of metamaterial unit cell into the design and simulation, extraction of absorption characteristics, and optimization towards performance enhancement. In this work, circular split ring resonator (SRR) is chosen as the metamaterial unit cell. The terahertz split ring resonator is designed using established techniques such as scaling and S-parameter retrieval. This SRR is then used as the basic element of a multi-layer structure, which together acts as an absorber. Then, a frequency tuning mechanism using MEMS switches is proposed. A PSO computational engine has been developed to integrate optimization algorithms with commercial EM solver. Further an adaptive tuning concept is introduced for multi-functional biomedical applications.

It is seen that conventional design techniques can be used to realize metamaterial absorbers with absorption greater than 95 %. Using the developed computational engine, the performance can be enhanced to near unity absorption with an optimized thickness. Further, this near unity absorption can be maintained for a tuning range of 0.5 THz using the MEMS-switches based technique presented in this work. A sensitivity analysis for taking into account fabrication and material tolerances is also presented.

In conclusion, a circular SRR-based active absorber has been designed and reported in this technical brief. The absorber shows near-unity absorption for a tuning range of 0.5 THz. The design of the absorber has been optimized for enhanced performance using particle swarm optimization, thereby validating the effectiveness of the methodologies presented in this document for the design of high performance, ultra-thin, metamaterial-based active terahertz devices.

Balamati Choudhury
Arya Menon
Rakesh Mohan Jha

Acknowledgments

We would like to thank Mr. Shyam Chetty, Director, CSIR-National Aerospace Laboratories, Bangalore for his permission and support to write this SpringerBrief.

We would also like to acknowledge valuable suggestions from our colleagues at the Centre for Electromagnetics, Dr. R.U. Nair, Dr. Hema Singh, Dr. Shiv Narayan, and Mr. K.S. Venu during the course of writing this book.

But for the concerted support and encouragement from Springer, especially the efforts of Suvira Srivastav, Associate Director, and Swati Meherishi, Senior Editor, Applied Sciences & Engineering, it would not have been possible to bring out this book within such a short span of time. We very much appreciate the continued support by Ms. Kamiya Khatter and Ms. Aparajita Singh of Springer towards bringing out this brief.

Contents

Active Terahertz Metamaterial for Biomedical Applications 1
1 Introduction 1
   1.1 Terahertz Radiation 2
   1.2 Motivation 3
2 Background Theory 4
   2.1 Terahertz Spectroscopy for Biomedical Applications. 5
   2.2 Metamaterials 6
   2.3 Tuning Mechanisms for Active Metamaterial Based
       Applications 11
   2.4 Metamaterial Absorbers 13
   2.5 Particle Swarm Optimization for Improvement
       of Design 14
3 Methodology. 18
   3.1 Metamaterial Design Procedure 18
   3.2 Design of Terahertz Metamaterials Using Scaling. 21
   3.3 Absorber Design 24
   3.4 Development of PSO-Based Computational Engine 27
4 Design and Result Analysis. 28
   4.1 Selection of Metamaterial Structure 28
   4.2 Design of Circular Split Ring Resonator 29
   4.3 Design of Absorber Using SRR 30
   4.4 Design of Active Absorber Array. 33
5 Conclusion 37
References 38

About the Book 43

Author Index 45

Subject Index 49

About the Authors

Dr. Balamati Choudhury is currently working as a scientist at Centre for Electromagnetics of CSIR-National Aerospace Laboratories, Bangalore, India since April 2008. She obtained her M.Tech. (ECE) degree in 2006 and Ph.D. (Engg.) degree in Microwave Engineering from Biju Patnaik University of Technology (BPUT), Rourkela, Orissa, India in 2013. During the period of 2006–2008, she was a Senior Lecturer in Department of Electronics and Communication at NIST, Orissa India. Her active areas of research interests are in the domain of soft computing techniques in electromagnetics, computational electromagnetics for aerospace applications and metamaterial design applications. She was also the recipient of the CSIR-NAL Young Scientist Award for the year 2013–2014 for her contribution to the area of Computational Electromagnetics for Aerospace Applications. She has authored and co-authored over 100 scientific research papers and technical reports including a book and three book chapters. Dr. Balamati is also an Assistant Professor of AcSIR, New Delhi.

Ms. Arya Menon has obtained her Bachelor of Engineering in Electronics and Communication Engineering from Manipal Institute of Technology, Manipal University. During the course of her studies, she interned at the Centre for Electromagnetics, CSIR-National Aerospace Laboratories (CSIR-NAL), and worked on soft computing for electromagnetics, conformal antennas, and metamaterials.

Dr. Rakesh Mohan Jha was Chief Scientist & Head, Centre for Electromagnetics, CSIR-National Aerospace Laboratories, Bangalore. Dr. Jha obtained a dual degree in BE (Hons.) EEE and M.Sc. (Hons.) Physics from BITS, Pilani (Raj.) India, in 1982. He obtained his Ph.D. (Engg.) degree from Department of Aerospace Engineering of Indian Institute of Science, Bangalore in 1989, in the area of computational electromagnetics for aerospace applications. Dr. Jha was a SERC (UK) Visiting Post-Doctoral Research Fellow at University of Oxford, Department of Engineering Science in 1991. He worked as an Alexander von Humboldt Fellow at the Institute for High-Frequency Techniques and Electronics of the University of Karlsruhe, Germany (1992–1993, 1997). He was awarded the Sir C.V. Raman

Award for Aerospace Engineering for the Year 1999. Dr. Jha was elected Fellow of INAE in 2010, for his contributions to the EM Applications to Aerospace Engineering. He was also the Fellow of IETE and Distinguished Fellow of ICCES. Dr. Jha has authored or co-authored several books, and more than five hundred scientific research papers and technical reports. He passed away during the production of this book of a cardiac arrest.

Abbreviations

BFO	Bacterial foraging algorithm
BST	Barium strontium titanate
CSRR	Circular split ring resonator
DES	Differential evolution strategy
ECA	Equivalent circuit analysis
ELC	Electric ring resonator
FEL	Free electron laser
FEM	Finite element methods
FIT	Finite integration techniques
FSS	Frequency selective surface
GA	Genetic algorithm
LHM	Left handed media
MOPSO	Multi-objective particle swarm optimization
NEP	Noise equivalent power
NN	Neural networks
PSO	Particle swarm optimization
SRR	Split ring resonator
THz-TDS	Terahertz time domain spectroscopy systems

List of Figures

Figure 1	Biomedical imaging system configuration	3
Figure 2	Diagram of a terahertz time domain spectroscopy system (THz-TDS)	6
Figure 3	Difference between a right-handed co-ordinate system and a left-handed co-ordinate system	7
Figure 4	Transmission line equivalents of **a** right-handed materials **b** left-handed materials	8
Figure 5	Common metamaterial designs **a** circular split ring resonator with linear wire **b** electric ring resonator **c** multi-band circular ring based metamaterial **d** square ring resonator	10
Figure 6	Position of all particles for all steps in the iterations	17
Figure 7	Variation of fitness function with iteration number	17
Figure 8	Diagram showing the E-fields in a slab of material with the given material parameters	19
Figure 9	Schematic of designed metamaterial. **a** Front view and **b** back view of metamaterial	22
Figure 10	Simulation results: **a** Permittivity and **b** Permeability of simulated metamaterial simulated (Fig. 9)	22
Figure 11	Simulation results. **a** Refractive index and **b** Impedance of simulated metamaterial (Fig. 9)	22
Figure 12	Permittivity (*left*) and permeability (*right*) of scaled metamaterial	24
Figure 13	Impedance (*left*) and refractive index (*right*) of scaled metamaterial	24
Figure 14	Multi-band metamaterial absorber.	25
Figure 15	Absorption in metamaterial absorber (Fig. 14)	26
Figure 16	E-fields in the structure at 3.96, 6.68 and 9.25 GHz respectively	26

Figure 17   Absorption characteristics of scaled absorber   27
Figure 18   Schematic of PSO-based computational engine   28
Figure 19   Circular split ring resonator for 2 GHz   29
Figure 20   Permeability of designed circular SRR for 2 GHz.
             The curve follows a Drude-Lorentz characteristic   30
Figure 21   Simulation results for 2 THz SRR. **a** Reflection, S21.
             b Real and imaginary parts of relative permeability   30
Figure 22   Schematic of optimized absorber   31
Figure 23   Absorption characteristic curve of absorber (Fig. 22)   32
Figure 24   PSO optimization: **a** variation of fitness function
             with respect to number of iterations, **b** absorption
             characteristics of optimized structure.   33
Figure 25   3 × 2 absorber array with CSRR unit cell   34
Figure 26   Rotation of inner ring by an angle θ with respect to outer
             ring   34
Figure 27   **a** Variation of fitness function with iterations
             for determination of angular rotation of inner rings
             in 3 × 2 absorber array, **b** absorption characteristics
             of absorber array   36
Figure 28   Implementation of adaptive tuning for absorber array   37
Figure 29   Variation of absorption considering different tolerances
             during fabrication.   37

List of Tables

Table 1   Dimensions of designed metamaterial   21

Table 2   Dimensions of metamaterial obtained after scaling.   23

Table 3   Dimensions of multi-band absorber (Fig. 14)   25

Table 4   Dimensions obtained after PSO optimization
of 2 THz absorber.   32

Active Terahertz Metamaterial for Biomedical Applications

Abstract Terahertz (THz) spectroscopy is gaining momentum as a tool for imaging in the field of biomedical engineering. This increase in popularity is due to the non-invasive, non-ionizing nature of terahertz radiation coupled with its propagation characteristics in water, which allows the operator to obtain high-contrast images of skin cancers, burns, etc. without detrimental effects. In order to tap this huge potential, researchers worldwide are aiming to build highly efficient biomedical imaging systems by introducing terahertz absorbers into biomedical detectors. The biggest challenge faced in the fulfilment of this objective is the lack of naturally occurring dielectrics, which is overcome with the use of artificially engineered resonant materials, viz. the metamaterials. This book describes such a metamaterial-based active absorber. The design has been optimized using particle swarm optimization (PSO), eventually resulting in an ultra-thin active terahertz absorber. The absorber shows near unity absorption for a tuning range of terahertz (THz) application.

Keywords Terahertz · Metamaterial · Soft computing · Ultra-thin active terahertz absorber · Active metamaterial

1 Introduction

Biomedical imaging for detection of detrimental skin diseases is one of the most exciting applications of terahertz technology; improvement in the performance of such systems would make them feasible from a commercial point of view. The performance of these terahertz imaging systems, terahertz time domain spectroscopy systems (THz-TDS), is dependent on the efficiency of its components such as detectors, etc. Performance of micro-bolometer detectors in THz-TDS can be significantly improved by adding an absorber to it. In this work, an attempt is made to design an absorber for this application. Due to the lack of natural dielectrics in the terahertz region, it becomes imperative that the absorber is designed using

© The Author(s) 2016
B. Choudhury et al., *Active Terahertz Metamaterial for Biomedical Applications*,
SpringerBriefs in Computational Electromagnetics,
DOI 10.1007/978-981-287-793-2_1

metamaterials, a class of artificially engineered materials. An attempt is also made to combat a drawback of metamaterial-based absorbers, viz. narrow bandwidth by introducing a frequency tuning mechanism into the design. Further, the active metamaterial-based absorber is optimized using a particle swarm optimization (PSO)-based computational engine in order to obtain heightened performance.

1.1 Terahertz Radiation

The frequency bands that lie within 0.1–10 THz is popularly called the terahertz region of the spectrum. Unlike its neighbors, microwave and IR, in the frequency spectrum, the development of applications in the terahertz spectrum has been slow. However, recent advances in science and technology have enabled researchers to explore the possible scope of terahertz as a viable and useful part of the spectrum. Current research has brought about multiple applications that can find relevance in human lives such as biomedical imaging, security, etc.

Terahertz radiation is difficult to produce. Consequently, efficient terahertz sources are hard to find. Coupled with this lack of sources, the primary reason for the slow developmental trends is attributed to the inability of most natural materials to exhibit dielectric properties toward incident terahertz radiation (Ferguson and Zhang 2009; Smith et al. 2004) resulting in a dearth of materials for fabrication of terahertz devices. Other properties of terahertz include non-ionizing nature, propagation along straight lines, and good penetrating power in non-conductors. Terahertz is also seen to attenuate highly in water as its absorption coefficient in water is reported to lie between 100 and 1000 cm^{-1} (Siegel 2004). This property limits the application of terahertz in long distance communication.

Significant advances in material science and design of terahertz sources have contributed to a sudden interest in this field. In fact, some contemporary terahertz applications take advantage of properties that were hitherto regarded as cons (Ferguson and Zhang 2009). Most of the current applications of terahertz such as those in space, security, biomedical fields are based on molecular sensing (deMaagt 2007). These applications exploit the high frequency characteristic of terahertz for imaging by increasing the resolution. While it can be argued that IR, MRI, and optical systems are capable of generating images of higher resolution, it has been observed that imaging in terahertz offers higher contrast, an important requirement for distinguishing substances (Siegel 2004). In fact, this contrast is instrumental in the application of terahertz for the detection of skin tumors. It has been observed that tumor cells contain more water than healthy tissue. Since terahertz attenuates differently in water, the terahertz radiation reflected off a tumor would be lesser than the surrounding regions (Kearney 2013). This property may also be used for the study of burns, etc. Molecular excitation using terahertz finds application in the study of protein states, DNA hybridization, detection of polymorphs in drugs, etc.

A schematic of a biomedical imaging system configuration is shown in Fig. 1.

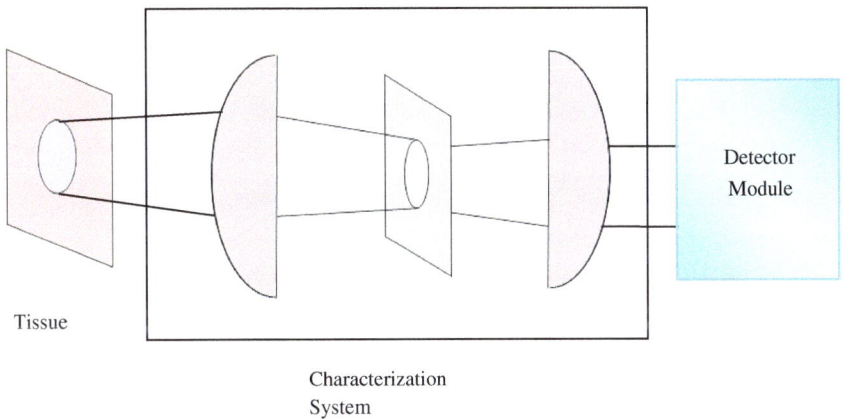

Tissue

Characterization
System

Fig. 1 Biomedical imaging system configuration

1.2 *Motivation*

Considering the multitude of possible applications of terahertz, it is natural for the electromagnetics community to look towards designing devices that work in this region of the spectrum, especially for applications in the biomedical field where this technology can help to detect dangerous diseases in the initial stages and possibly save human lives! However, design of terahertz devices is not an easy task as the designer has to combat multiple challenges. Therefore, successfully developing procedures that can be used to design terahertz devices for biomedical applications is the need of the hour!

Biomedical Imaging: As mentioned in the previous section, terahertz technology holds great potential for the detection of many life-threatening diseases. However, the true necessity of terahertz imaging is understood only when the drawbacks of contemporary imaging equipment are discussed.

Drawbacks of Contemporary Imaging Systems: X-rays and MRI are the two most popular imaging systems in the medical field. X-rays have high penetration power and are generally used to study bones. In addition, prolonged exposure to X-rays has been reported to be extremely harmful. On the other hand, exposure to MRI is not detrimental and this technology is used to study internal organs. However, none of these two technologies are used for studying skin-related diseases such as carcinoma of the skin, etc.

Advantages of Terahertz Instruments: Terahertz radiation has proven to be non-ionizing and hence it is not harmful to human beings. Additionally, properties like low penetration depth and ability to generate high contrast images make it suitable for imaging of the skin.

Challenges of Terahertz Technology: It has been mentioned that development of applications in terahertz has been slow. This trend can be attributed to the various challenges faced by researchers while designing devices for terahertz.

Material Issues: Naturally available materials fail to display dielectric properties when terahertz radiation is incident on them. It is observed that in most cases, the relative permittivity becomes equal to 1. This calls the need for novel, artificially engineered material called metamaterial, which show dielectric properties.

Design Issues: Metamaterials are sub-wavelength structures, and analysis of these materials is carried out using complicated methods like FDTD, equivalent circuit analysis (ECA), etc. Further, mathematical expressions that relate the resonant frequency of an absorber with its dimensions are not easily available. Therefore, metamaterial designs are often accompanied by cumbersome, iterative simulations.

Fabrication Issues: Two issues must be discussed during the design of devices operating in the terahertz region: (a) identification of fabrication techniques that can realize the required terahertz metamaterial feature size, and (b) inclusion of modifications into chosen fabrication technique for mass production.

Characterization Issues: Contemporary THz sources, such as gyrotrons, resonant tunneling diodes, and backward wave diodes, generate radiation with power levels lying in the μW–mW range. Therefore, receivers require very high sensitivity. As a result, research must be conducted in order to develop efficient, high power sources, and highly sensitive detectors.

In this work, the possible solutions to the above-mentioned challenges in terahertz technology will be explored. The focus of this work will be on design of high performance, artificially engineered material-based terahertz absorbers with adaptive tuning mechanism. This absorber can be used as a component of micro-bolometer-based terahertz detectors in biomedical imaging systems, thereby improving the performance of these detectors.

Metamaterials Instead of Dielectrics: Design of artificially engineered materials, called metamaterials, in order to combat lack of naturally occurring dielectrics in the terahertz spectrum and development of terahertz absorber using metamaterials for application in biomedical imaging.

High Performance Design: Development of soft computing based optimization tools in order to design high performance EM structures and usage of this technique to optimize metamaterial absorber (Choudhury et al. 2012).

Adaptive Tuning: Improve the frequency of operation of the designed metamaterial absorber by (a) integrating a frequency tuning mechanism with the design, and (b) implementing tuning mechanism in order to design an absorber array with adaptive tuning capability.

2 Background Theory

This section provides a basic understanding about all the concepts used in this work. First, this section covers the basic design of systems used for terahertz spectroscopy. The need for an absorber in the detector is then discussed. As established in the introduction, the absorber proposed in this work is a metamaterial absorber and hence, the theory behind metamaterials will be presented. The limitation in the bandwidth of metamaterials is then brought out and consequently, the need for incorporation of tuning mechanisms in order to improve the versatility of the absorber is presented. A comprehensive literature survey is then conducted in order to identify the contemporary tuning mechanisms for terahertz metamaterials. Following this, the design of metamaterial absorbers is presented. In the last part of this section, a need for optimization of absorber design using soft computing is discussed along with the description of PSO.

2.1 Terahertz Spectroscopy for Biomedical Applications

Terahertz technology finds immense applications in the field of medicine and biology. Since terahertz is ionizing in nature, it can be used to study tissue in situ. Consequently, terahertz offers a non-invasive technique to characterize human tissue.

Terahertz instruments may be broadly classified into two types: passive instruments and active instruments. Passive instruments generate incoherent terahertz radiation. This incoherent nature leads to inaccurate imaging and hence, passive terahertz instruments are not used for biomedical applications. However, such instruments may be used in the design of security systems. Certain companies have implemented passive terahertz imagining technology for detecting forbidden and concealed items such as knives, etc.

On the other hand, active instruments produce terahertz radiation that is coherent in nature; either as continuous or pulsed waveforms. Terahertz time domain spectroscopy (THz-TDS) and free electron laser (FEL) are two popular instruments that belong to the category of active terahertz instruments (Humphreys et al. 2004). FEL generates high power, large bandwidth, coherent radiation. This is achieved by passing an electron beam through sets of oppositely polarized magnets. The electrons oscillate due to the forces exerted by the magnets and produce terahertz radiation. The drawback of this technology is that it is large in size and complex to construct. Additionally, FEL instruments are also expensive.

These drawbacks have prompted many researchers to consider THz-TDS as a viable option for imaging purposes. A typical THz-TDS system is shown in Fig. 2. Short pulses of time period 90 fs are generated by a laser. This pulse is then split. One portion of the split pulse is made to fall on a biased GaAs wafer, which generates pulses of terahertz radiation. At the detector, the other portion of the laser pulse (after being passed onto a delay line) is used to sample the terahertz radiation reflected off/transmitted through the sample at the detector.

Detectors and Need for Absorbers: Detectors are of the most important component of the THz-TDS system. These detectors have to be compact, inexpensive, and highly efficient. Typical THz-TDS systems employ uncooled micro-bolometers (Kearney 2013). One of the most common detectors used in imaging, especially thermal imaging, is the uncooled micro-bolometer detector (Kearney 2013). These

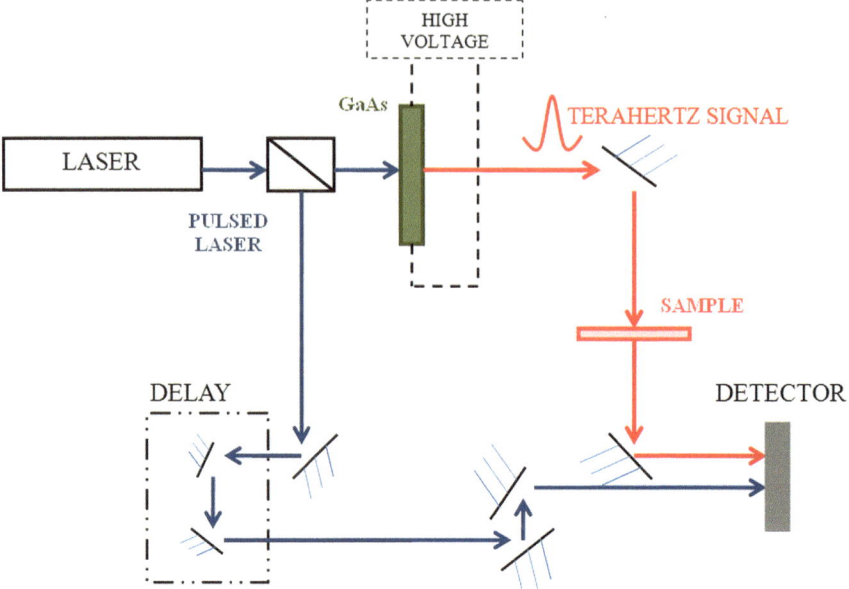

Fig. 2 Diagram of a terahertz time domain spectroscopy system (THz-TDS)

detectors exhibit changes in their electrical resistance when exposed to heat (IR radiation). The fact that the terahertz band is adjacent to the IR band means that the same detector may be used for detection of terahertz as well, but with degradation in performance. Moreover, the noise equivalent power or NEP in the terahertz range is about 300 pW/Hz$^{1/2}$ compared to 14 pW/Hz$^{1/2}$ for IR. Though the detectors demonstrate poor responsivity in terahertz, their performance can be enhanced substantially by using absorbers.

Traditional absorber designs make use of multiple layers of dielectric spacers and resistive sheets. However, as mentioned earlier, naturally occurring materials do not exhibit dielectric properties in the terahertz band. To overcome this, metamaterials or artificially engineered materials that are material-based absorbers are incorporated into bolometers for increasing the sensitivity of the detectors in the terahertz band.

2.2   Metamaterials

The term metamaterial was coined by Rodger M. Wasler of University of Texas at Austin. The root of this word, '*meta*', is Greek and means beyond. Indeed, these materials exhibit certain characteristics that are not seen in materials found in nature —the most exciting one being the ability to display negative refractive index.

Though the science of metamaterials is relatively new, the possibility of the existence of doubly negative materials (those that display negative permeability and negative permittivity simultaneously) was predicted by Victor Veselago in 1968 in a Russian publication.

Left-Handed Materials: Doubly negative materials are also called *left-handed media* (LHM). When both the permittivity and permeability become negative, the refractive index also becomes negative. Consequently, wave propagation in these materials follows left-handed co-ordinate frame as shown in Fig. 3. In a left-handed frame, the positive rotation is clockwise about the axis of rotation. The effect of this change in co-ordinate system is profound, i.e., mathematically, it is seen that the Poynting and group velocity vectors have opposite directions! In simple terms, this means that the direction of phase propagation is opposite to the direction of propagation of power! It is also important to note that LHM is necessarily dispersive in order to meet entropy conditions. This means that either the permittivity or permeability (or both) varies as a function of frequency (Caloz 2006). Other properties are as follows:

LHM include inverted Doppler effect, inverted Čerenkov effect, artificial magnetism, reversal of Snell's law, etc.

Transmission line equivalent of LHM: Right-handed materials can be represented as a lossless transmission line as shown in Fig. 4a. Using this equivalent transmission line, it becomes easier to arrive at characteristic impedance, propagation constant, etc., of this material. An LHM can be modeled by taking the dual of the transmission line of the RHM and is shown in Fig. 4b (Caloz 2006). The impedances corresponding to the capacitance and inductance are given by Eqs. (1) and (2), respectively.

$$Z = \frac{1}{j\omega C} \tag{1}$$

$$Y = \frac{1}{j\omega L} \tag{2}$$

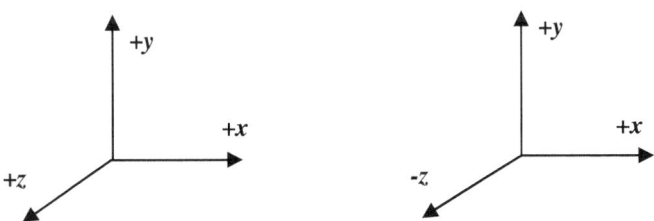

Right-handed co-ordinate system   Left-handed co-ordinate system

Fig. 3 Difference between a right-handed co-ordinate system and a left-handed co-ordinate system

Fig. 4 Transmission line equivalents of **a** right-handed materials **b** left-handed materials

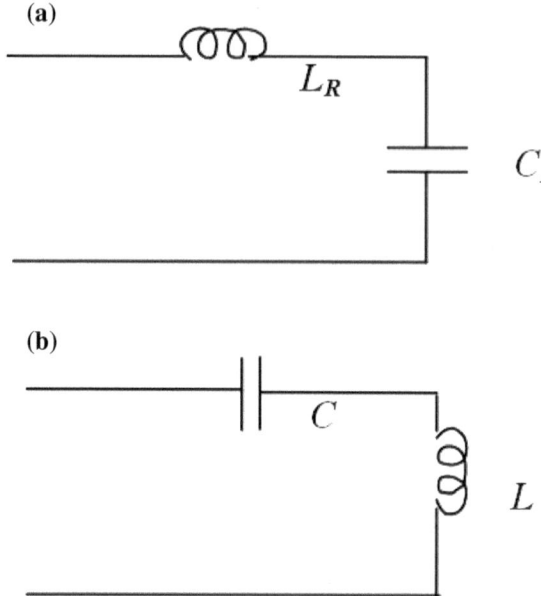

The propagation constant, γ for an angular frequency ω is then given as

$$\gamma = j\beta = \sqrt{ZY} = \frac{1}{j\omega\sqrt{LC}} = \frac{-j}{\omega\sqrt{LC}} \tag{3}$$

where, β is the phase constant, and attenuation constant, $\alpha = 0$ as the analysis is carried out for lossless transmission lines. Clearly, β is a negative value.

The phase velocity v_p and group velocity v_g are then given as

$$v_p = \frac{\omega}{\beta} = -\omega^2\sqrt{LC} < 0 \tag{4}$$

$$v_g = \frac{d\omega}{d\beta} > 0 \tag{5}$$

Hence, the phase and group velocities are antiparallel, which shows that the direction of propagation of phase in this material is opposite to the direction in which power is transferred. While this seems to be slightly troubling, it must be noted that the phase velocity gives the propagation of a perturbation and hence by becoming negative, it does not violate the laws of physics. On the other hand, the group velocity can never become negative as this would lead to a non-causal system with power being transferred into the source.

Definition of Metamaterials: While LHM are the most popular form of metamaterials, it must be noted that the definition of metamaterial can be extended to include other types of material also. One popular definition of metamaterials is given as:

> A metamaterial is a macroscopic composite of periodic/non-periodic structure, whose function is due to both cellular architecture and chemical composition. (Cui et al. 2010)

This definition clearly states that the material parameters of a metamaterial, viz. the permittivity and permeability can be engineered by changing its structural and material properties. At this juncture it should be stated here that typical metamaterial structures consists of metallic patterns etched onto a dielectric substrate. Different patterns have been reported in literature like electric ring resonators with wires (Yang et al. 2013a, b), graphene-based films (He 2013), stacked rings (Kung and Kim 2013), combination of FSS, I shaped rods, etc. (Hokmabadi et al. 2013). The dimensions of these metallic structures as well as the material and height of the substrate significantly affect the resonant frequency. The cellular size must be smaller than or equal to subwavelength (Cui et al. 2010). This corresponds to a cell size, p lesser than a quarter of a wavelength. When $p = \lambda/4$, the value is called *effective-homogeneity limit* and ensures that refraction dominates scattering/diffraction when wave propagates in a metamaterial.

Common Metamaterial Designs: Metamaterials can be classified into two types

- Resonant metamaterials
- Non-resonant metamaterials

Resonant Metamaterials: Metamaterials that display a dynamic range of permittivity or permeability in the neighborhood of the resonant frequency are classified as resonant metamaterials. This means that a large change in any of the material parameters is observed when the frequency is changed by a small amount. The plot of permittivity *versus* frequency and permeability versus frequency are seen to follow a Drude-Lorentz relationship as it displays a particular response (Cui et al. 2010). However, narrow bandwidth of μ and ε as well as high loss near the resonant frequency is the biggest disadvantages of resonant metamaterials. Split ring resonators, ELC's, etc., are examples of resonant metamaterials.

Non-Resonant Metamaterials: Metamaterials, whose μ and ε vary slowly with respect to frequency are called non-resonant metamaterials. The loss in these materials is very small. Ring-like structures are the most popular form of non-resonant metamaterials. Therefore, observing the material parameters, viz. ε and μ provides insight into the performance of a metamaterial. However, the challenge faced is to obtain these values. Conventionally, metamaterials are simulated using software using finite element methods (FEM) or finite integration techniques (FIT). These software enable the designer to study the S-parameters of the metamaterial and in order to obtain the effective material parameters, S-parameter retrieval technique is used.

As mentioned earlier, metamaterials are periodic/aperiodic structure composed of metallic strips etched onto dielectric substrates. The geometry of the metallic layer as well as the properties of the dielectric is responsible for the resonant characteristics of the metamaterial. Figure 5 shows some of the commonly used

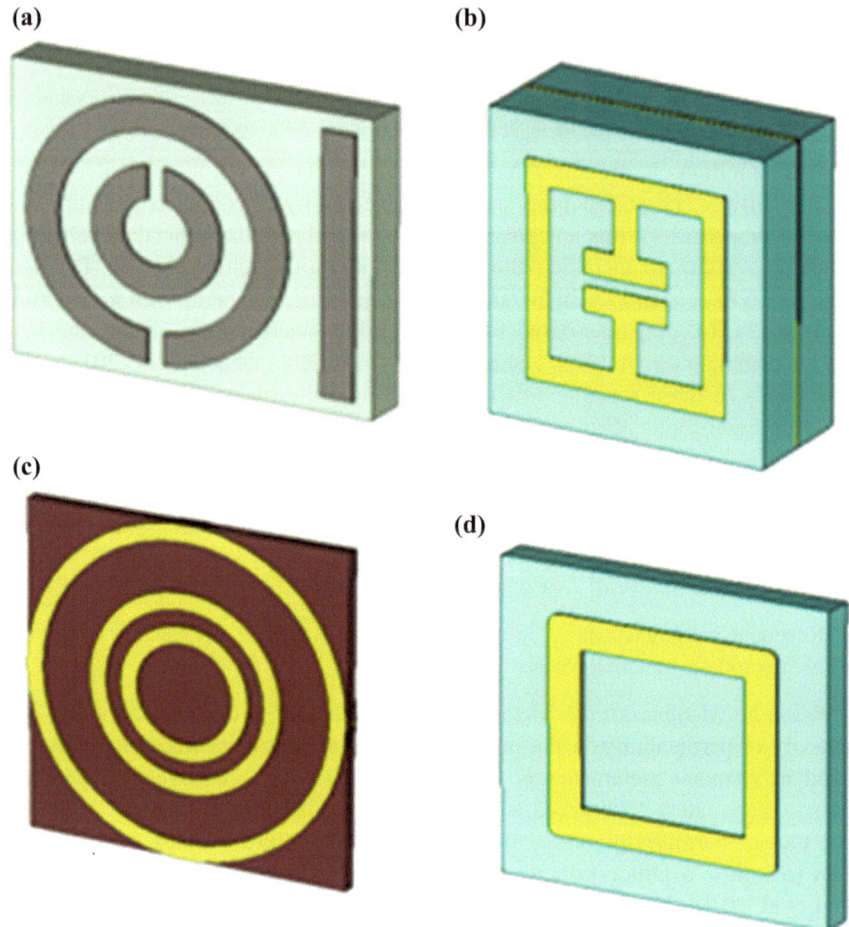

Fig. 5 Common metamaterial designs **a** circular split ring resonator with linear wire **b** electric ring resonator **c** multi-band circular ring based metamaterial **d** square ring resonator

metamaterial designs. Figure 5a is a circular split ring resonator (SRR) with a linear wire. The SRR is responsible for the magnetic response of the metamaterial, while the linear wire is responsible for the electric response. Figure 5b shows an electric-ring resonator (ELC). The resonant characteristics of the ELC are similar to that of the SRR. Figure 5c, d are non-resonant metamaterials. Due to multiple rings, the metamaterial given in Fig. 5c resonates three frequencies. These designs have been explored by many research groups around the world.

Moser et al. (2005) designed and fabricated an electromagnetic metamaterial using micro-fabrication technique. The structure was based on a rod-split ring resonant structure proposed by Pendry. The response of this structure was numerically simulated and was further verified using far IR transmission spectroscopy. It was shown that the composite structure behaves as a metamaterial in the frequency range of 1–1.27 THz (Moser et al. 2005). The authors proposed that this micro-fabrication technique can be extended to fabricate metamaterial structures that operate in the near IR range of frequencies.

Alaee et al. (2012) designed and obtained the optical response of a perfect absorber on a curved surface. A planar geometry that consists of parallel metallic nanowires above a metallic ground plane separated by a dielectric spacer was first designed and its absorption for different angles of incidence at a resonant frequency of 232 THz was studied. For low angle of incidence, the absorption was unity. As angle of incidence increases, the resonant frequency is shifted to higher frequencies. This planar perfect absorber was then wrapped around a cylindrical dielectric. The application of this absorber for suppression of spurious back-scattered light, and as an optical absorber was demonstrated using the response. Suggestions to increase the bandwidth of the absorber were also mentioned.

Alves et al. (2012) realized a narrowband (1 THz) emitter based on metamaterial structures at specific frequencies in the 4–8 THz region. The structure consisted of periodically repeated Al squares separated from an Al ground plane using silicon dioxide. The resonating frequency of the metamaterial was dependant on the size of the Al square. On heating the metamaterial to 400 °C, the emission spectrum was obtained with emissivity approaching unity at the resonant frequency. Emitters resonating at multiple frequencies were also designed and characterized by using Al squares of different sizes.

Li et al. (2012) designed, simulated, fabricated, and tested a dual-band terahertz metamaterial. The metamaterial consists of cruciform and square structures. As a result of this configuration, the negative permittivity was observed in the following bands: 378–500 GHz (due to the cruciform structures) and 626–677 GHz (due to the squares).

Huang et al. (2012) used *continuous ant colony optimization* (CACO) and *differential evolution strategy* (DES) in order to optimize a broadband terahertz absorber. After the implementation of the algorithms, optimized dimensions of the structure were obtained such that the band of frequencies over which the absorption was greater than 90 % increased to 0.872–0.975 THz from 0.885 to 0.965 THz.

2.3   Tuning Mechanisms for Active Metamaterial Based Applications

Biomedical applications require operation over a wide frequency range. However, most metamaterial structures, like the ones mentioned in the previous section, are frequency specific. Therefore, a need for mechanisms for changing the frequency arises which calls for an investigation into the different techniques that can enable such frequency tuning. Numerous examples of such tuning mechanisms can be found in literature especially for metamaterials designed for applications in the IR and microwave frequency bands.

An example of X-band, active, micro-split ring activated metamaterial is described by Ekmekci et al. (2009). The shorting of these mirco-splits (through MEMS switches) changed the resonant frequency of the metamaterial.

Similarly, the tuning of frequency by modifying gaps in split ring resonators using varactor diodes was proposed by Shadrivov et al. (2006). Pratibha et al. (2009) proposed a metamaterial structure by dispersing gold nanoparticles in liquid crystal. The variation in resonant frequency was achieved by varying the orientation of the particles within the medium.

From a preliminary study of active metamaterials, it can be concluded that tuning mechanisms can be classified into four different types viz. MEMS based, electrical actual, photoexcitation, and thermal actuation. A literature survey classifying THz tuning mechanisms into any of these four categories has been given below.

MEMS Based: Cantilever beams are a popular mechanism for tuning in the terahertz region. Tao et al. (2008) and Ozbey et al. (2011) implemented the cantilever structure into split ring resonators. The frequency of operation could be controlled by flexing the beam. While the principle behind the operation was the same, the two research groups with different methods were in order to achieve the movement of the beam. Another popular technique involves movement of parts of a metamaterial in order to either introduce anisotropy (Zhu et al. 2012) or change the effective capacitance of the metamaterial structure (Yang et al. 2013a, b). Other examples of multi-layer metamaterials inherently functioning as cantilevers (Vendik et al. 2012) and movement of structural elements in order to control polarization of incident terahertz radiation is also available (Zhang et al. 2013).

Photo Excitation: This tuning technique takes advantage of excitation of electrons in certain material when exposed to light of a specific wavelength. Padilla et al. (2006) used this technique to vary the properties of a Gallium Arsenide (GaAs) substrate on which a split ring resonator was etched. Other research groups have also demonstrated that photoexcitation may be used in order to control gap capacitances in resonant structures by changing the gap length. When light is incident on a doped gap, the properties of the gap changes. The resonant characteristic of the metamaterial structure changes in this process (Kozlov et al. 2011; Chowdhuryet al. 2012).

Electrical Actuation: Similar to photo excitation, electrical actuation involves varying the properties of materials by providing different bias voltages to it. Paul et al. (2009) used a assembly of gold crosses placed on a biased, n-doped GaAs substrate, in order to design a terahertz modulator. Kowerdziej et al. (2012) also used the same technique in order to change the properties of a nematic liquid crystal. The LC was sandwiched between two omega-shaped resonators in order to develop a THz transducer.

Thermal Actuation: Thermal actuation exploits the sensitivity of some materials and their properties to heat. Several research teams have successfully developed active metamaterials using this technology. Němec et al. (2009) fabricated a terahertz metamaterial made of high permittivity $SrTiO_3$ (STO) rods placed at regular intervals. The dielectric properties of these rods could be varied by changing the temperature. Variation in the dielectric properties resulted in variation of the resonant frequency.

2.4 Metamaterial Absorbers

From Sect. 2.1, it is clear that the performance of a terahertz time domain spectroscopy system (THz-TDS) is dependent on the quality of the detector. The most popular kind of detector, the micro-bolometer detector is usually designed for operation in the IR range of the frequency spectrum. The application of these detectors in the terahertz band is possible by using absorbers to improve the total noise power equivalent.

In this section, the drawbacks of conventional absorbers in the terahertz domain are discussed, followed by a discussion on metamaterial absorbers.

Need for Metamaterial Absorbers: Conventional absorbers are multilayered in design and often consist of dielectrics and resistive sheets stacked one behind the other such as Jaumann absorber, Salisbury Screen, Dallenbach layer, etc. (Vinoy and Jha 1996). However, as mentioned in Sect. 1.2, material ceases to show dielectric properties when exposed to terahertz radiation. Therefore, these conventional absorber designs cannot be used in the terahertz domain. As a result, one has to explore the unique properties of metamaterials for the design of terahertz absorbers.

Metamaterial Absorber Design Procedure: As the name suggests, an absorber is a structure that absorbs EM energy incident on it. Metamaterial absorbers were first proposed by N.I. Landy in 2008. Absorption of both the electric and magnetic energy is essential for improved performance. This requires tuning of material properties viz. the permittivity and permeability in order to make the impedance of the absorber equal to that of free space (Butler 2012). The impedance of free space is given as

$$Z_o = \sqrt{\frac{\mu_o}{\varepsilon_o}} \tag{6}$$

Similarly, the impedance of the absorber is given as

$$Z = \sqrt{\frac{\mu}{\varepsilon}} = \sqrt{\frac{\mu_o \mu_r}{\varepsilon_o \varepsilon_r}} \tag{7}$$

Therefore in order to make $Z = Z_o$, the relative permittivity, ε_r must be made equal to the relative permeability μ_r. Physically, 100 % absorption (A) is achieved when the transmission (T) and reflection (R) through the metamaterial is zero. This is easily understood through equation

$$A + R + T = 1 \tag{8}$$

In order to realize the condition of zero transmission, the metamaterial layer is often backed by a ground plane.

2.5   *Particle Swarm Optimization for Improvement of Design*

The design of metamaterial absorbers is a complicated procedure and requires determination of structural parameters of the metamaterial layer, as well as the thickness of the dielectric substrate for resonance at a particular frequency. As a result, a good design is obtained when the aforementioned parameters are optimized. This brings about the need for the implementation of various optimization strategies in metamaterial design.

In this section, soft computing is presented as a feasible optimization technique for metamaterial design (Choudhury et al. 2012). The advantages of soft computing over traditional computing techniques are discussed, and the implementation of soft computing in electromagnetic problems is presented. Following this, a specific soft computing technique viz. PSO is discussed. PSO will be used as the optimization tool in this work.

Introduction to Soft Computing: Soft computing is not a single technique, but a group of computational techniques comprising of artificial intelligence, machine learning, and evolutionary techniques used for analysis and modeling of complex phenomenon where low-cost analytical formulations does not exist. The term soft computing was first coined by Zadeh in (1992). Soft computing techniques are characterized by their ability to arrive at accurate, low-cost, robust solutions despite approximations and uncertainty in their design. Some of the commonly used soft computing techniques include neural networks (NN) and fuzzy logic, genetic algorithm (GA), particle swarm optimization (PSO), bacterial foraging algorithm (BFO), etc.

The unique properties of soft computing methods are:

- Soft computing methods do not require extensive mathematical formulation of the problem. Thus, the requirement on the necessity of exclusive domain-specific knowledge can be reduced.
- These tools can handle many variables simultaneously.
- For optimization problems, the solutions can be prevented from falling into local minima by using global optimization strategies.
- These methods yield cost-effective solutions to the user. Further, hybridization of soft computing methods may reduce the conventional simulation procedures up to some extent.
- These methods are adaptive and can be scalable.

Abundant literature is available for implementation of genetic algorithm and micro-genetic algorithm for RAM optimization. Chakravarty et al. (2001) used the same in order to design a frequency selective surface (FSS) based broadband microwave absorber. The work also shows that implementation of micro-genetic algorithm over genetic algorithm considerably speeds up the computation time. The algorithm was designed to simultaneously select the best materials and their thicknesses as well as vary the structural parameters of the FSS for optimized performance.

Guodos and Sahalos (2006) designed and optimized multi-layer planar absorptive coatings using multi-objective particle swarm optimization (MOPSO). The designed coating showed wide band and wide angle characteristics. Further, a comparison between MOPSO and multi-objective GA revealed that MOPSO was more efficient and required lesser computing time.

Therefore, it is well established that soft computing can be used as an effective tool for electromagnetic designs. As mentioned above, compared to conventional soft computing techniques like neural networks and genetic algorithm, PSO is found to require lesser memory, is simple and efficient, and has inherent techniques to prevent the algorithm from falling into a local minimum. As a result, in this work, PSO is chosen as the optimizing tool for absorber design.

Particle Swarm Optimization: PSO is a soft computing technique that emulates swarm intelligence. This algorithm was first proposed by Kennedy and Eberhart in (1995). The working of this algorithm is best described by citing the example of a swarm of bees trying to locate the point with maximum density of flowers in a field (Jin and Samii 2007). Initially, the population of bees have no knowledge about the location of the point with maximum flower density. Therefore, they randomly disperse themselves throughout the field. Each bee then determines the density of flowers at the point it is, and exchanges its findings with all other bees. It then moves in the direction of the bee which has reported the highest density of flowers. In essence, the movement of each bee is influenced by two factors: its own best value and the best value in the swarm. PSO models this behavior in the form of a computer algorithm. The "density of flowers" is replaced by a problem-specific objective function. Conventionally, the best location is the one which results in minimum value of the objective function.

The terminology used in PSO is given below:

Particle: Potential solution in the solution space (analogous to a bee)

Fitness: Result of evaluation of the problem's objective function for the particles co-ordinates

Personal Best (pbest): Best value of fitness obtained for each particle

Global Best (gbest): Best value of fitness among all particles in all iterations

Algorithm for PSO: The following is the algorithm used for implementing PSO:

Step 1: *Initialize random positions and velocities for each particle*
         Particles are assigned a random position within the N-dimensional solution space [xMin, xMax]. Here, it is assumed that the range for each dimension is the same. However, this may not be the same for few

problems and hence, must be assigned properly. Each particle is then assigned a velocity lying within the range [vMin, vMax]. The values for vMin and vMax are calculated as follows

$$vMin = -0.2(xMax - xMin) \tag{9}$$

$$vMax = 0.2(xMax - xMin) \tag{10}$$

Step 2 *Evaluate the fitness and determine pbest, gbest*:
Evaluate the fitness function (objective function) f for the current position of each particle. Initially, these evaluated values themselves correspond to pbest. For subsequent iterations, pbest is updated by the calculated fitness if it is lesser than the current pbest. The value of gbest is taken as the minimum of all pbest and is updated only if a fitness value lesser than the current gbest is obtained.

Step 3 *Update Position and Velocity*:
Depending on the values of pbest and gbest, a position (x) and velocity (v) update for each particle is carried out using the following equations:

$$v = w \times v + c_1 r_1 (\text{pbest} - x) + c_2 r_2 (\text{gbest} - x) \tag{11}$$

$$x = x + v \tag{12}$$

where, w is called the inertial constant, c_1 is called the cognitive constant, and c_2 is called social constant. From Eqs. (11) and (12), it is seen that the movement of particles is described using Newtonian mechanics. For best results, the value of w is decremented from 1, from the start of the algorithm to 0.4 at the end of the algorithm (Jin and Samii 2007). Further, research has found that a value of 0.5 for the social and cognitive parameters increase the rate of convergence in the algorithm (Parsopoulos and Vrahatis 2002).

Step 4 *Run loop*:
Return to Step 2 and perform each step for N_t times.

Testing of developed PSO code A Matlab code for single objective PSO was written using the algorithm given in the previous section. The fitness function was chosen to be

$$y = 1 - \frac{\sin[\pi(x_1 - 10)]}{\pi(x_1 - 10)} \frac{\sin[\pi(x_2 - 10)]}{\pi(x_2 - 10)} \tag{13}$$

Twenty particles and 35 iterations were used in the code. The value of x_1 and x_2 taken by all the particles for the entire duration of the run of the algorithm is given in Fig. 6. Toward the end of the iterations, the particles clustered around (10, 10), which is in fact the solution to Eq. (13). The variation of the fitness function as the algorithm progresses is given in Fig. 7.

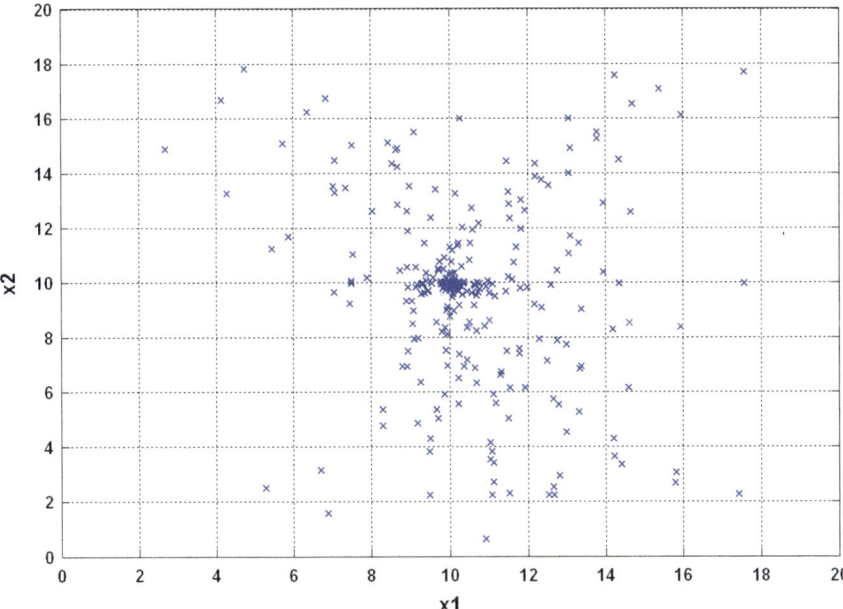

Fig. 6 Position of all particles for all steps in the iterations

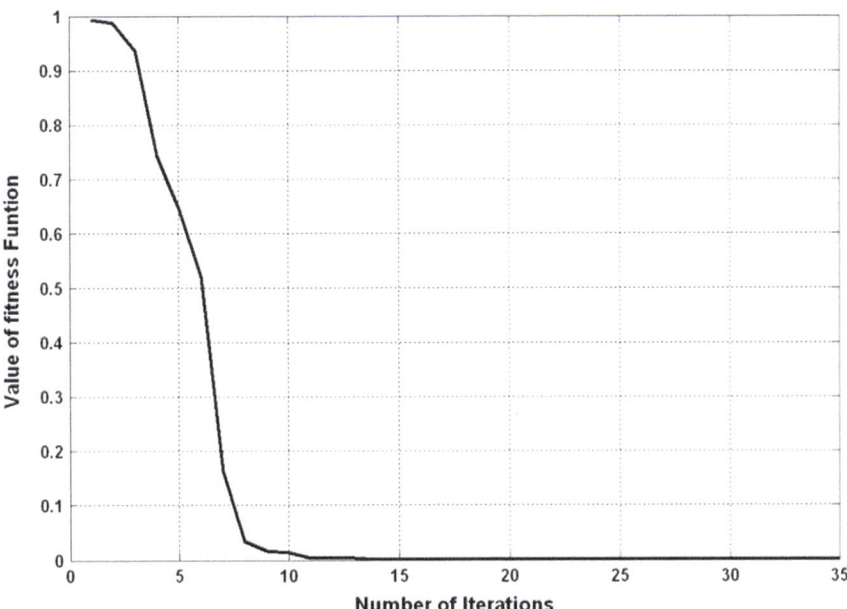

Fig. 7 Variation of fitness function with iteration number

3   Methodology

Due to the lack of availability of natural dielectrics, terahertz devices are designed using metamaterials. The design of metamaterials is a complicated procedure and requires the aid of commercial EM solvers for design verification. This section discusses the techniques used in order to design metamaterials for this work. The preliminary design strategies used are S-parameter retrieval method and scaling. The theory behind both these strategies is presented along with verification of these methods through validation of previously reported work. The feasibility of the scaling technique for metamaterial-based absorber design is then discussed.

While these strategies are suitable for preliminary designs, other techniques must be explored in order to realize high performance metamaterial absorbers. As mentioned in the previous two sections, PSO is a soft computing technique that has the potential to arrive at robust, quick, and accurate solutions for metamaterial-based problem statements. In this section, an overview of a PSO-based computational engine that integrates the PSO algorithm with a commercially available EM simulator is presented.

3.1   Metamaterial Design Procedure

Verification of metamaterial design is conducted by studying its permittivity and permeability. Structural parameters are modified in order to achieve the desired value of permittivity. Therefore, techniques to obtain these parameters from simulations must be explored. From literature, it is seen that the most popular extraction technique is the S-parameter retrieval method.

S-parameter Retrieval Method: The Nicolson Ross-Weir S-parameter retrieval method is one of the most commonly used parameter retrieval methods for metamaterial samples, illuminated by plane wave in the normal direction (Arslanagić et al. 2013). The ε and μ extracted using this technique give the equivalent material parameters, i.e., a homogeneous slab of material with this extracted permittivity and permeability shows the same S-parameters as that of the metamaterial in the design. However, it must be noticed that macroscopic field inside the metamaterial need not correspond to the one inside the homogeneous slab. Further, the equivalent material parameters can be equated to the effective material parameters when the periodicity of unit cells in a metamaterial array is much smaller than the wavelength λ.

However, this method does not provide a unique solution for the values of permittivity (ε) and permeability (μ). Mathematically, this ambiguity is seen as branches of a complex logarithmic function, the explanation for which is given using Bloch state physics. But, techniques to compensate for the ambiguity have been explored by different research groups and are available in literature. Another drawback with this method occurs when the sample thickness is an integral multiple of $\lambda/2$; non-physical phenomena have been observed. Consider Fig. 8, a plane wave

Fig. 8 Diagram showing the E-fields in a slab of material with the given material parameters

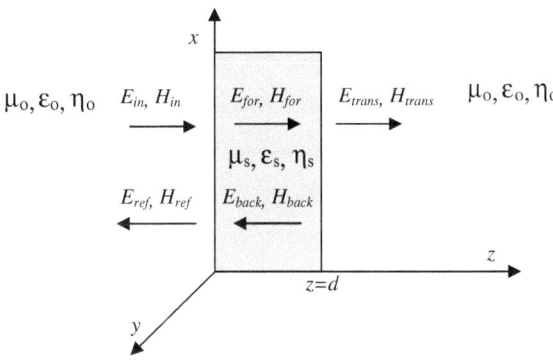

propagating in the z direction is incident on a slab of material with permittivity ε_s and permeability μ_s. The impedance of the material is given as η_s. Let, μ_o, ε_o, and η_o represent the permeability, permittivity, and impedance of free space. Further, let the E-field of the incident wave be in the x direction.

The fields can then be written as

$$E_{\text{in}}(z) = E_{\text{in}}e^{-jk_sz}\hat{x} \tag{14}$$

$$H_{\text{in}}(z) = \frac{E_{\text{in}}}{\eta_o} e^{-jk_sz}\hat{y} \tag{15}$$

$$E_{\text{ref}}(z) = E_{\text{ref}}e^{jk_sz}\hat{x} \tag{16}$$

$$H_{\text{ref}}(z) = \frac{E_{\text{ref}}}{\eta_o} e^{jk_sz}\hat{y} \tag{17}$$

The other fields can also be expressed in a similar manner. It is seen that the thickness of the slab is d (Arslanagić et al. 2013). Enforcing boundary conditions at $z = 0$ and $z = d$, we obtain

$$E_{\text{ref}} = \frac{E_{\text{in}}(1 - z^2)(\eta_o^2 - \eta_s^2)}{D} \tag{18}$$

$$E_{\text{back}} = \frac{E_{\text{in}}z^2\eta_s(\eta_o - \eta_s)}{D} \tag{19}$$

$$E_{\text{for}} = \frac{2E_{\text{in}}\eta_s(\eta_o + \eta_s)}{D} \tag{20}$$

$$E_{\text{trans}} = \frac{4E_{\text{in}}\eta_o\eta_s ze^{jk_od}}{D} \tag{21}$$

where,

$$D = (\eta_o + \eta_s)^2 - (\eta_s - \eta_0)^2 z^2 \qquad (22)$$

$$z = e^{-jk_s d} \quad k_s = \omega \varepsilon_s \mu_s \qquad (23)$$

From these, the S-parameters are derived using the Nicolson-Ross-Weir method (Arslanagić et al. 2013).

$$S_{11} = \frac{(1 - z^2)(\eta_s^2 - \eta_o^2)}{(\eta_o + \eta_s)^2 - (\eta_s - \eta_o)^2 z^2} \qquad (24)$$

$$S_{21} = \frac{4\eta_o \eta_s z}{(\eta_o + \eta_s)^2 - (\eta_s - \eta_o)^2 z^2} \qquad (25)$$

It is seen that there are multiple sets of ε and μ that give the same S-parameters. Let these sets be indexed using integer p, i.e., different values of p give different parameters such that

$$\frac{\mu_{sp}}{\varepsilon_{sp}} = \frac{\mu_{s1}}{\varepsilon_{s1}} \qquad (26)$$

where, (μ, ε) denotes a set of slab metamaterial properties.

The impedance of the slab can then be calculated by rearranging and solving the above equations (Parsopoulos and Vrahatis 2002).

$$\eta_s = \pm \eta_o \sqrt{\frac{(S_{11} + 1)^2 - S_{21}^2}{(S_{11} - 1)^2 - S_{21}^2}} \qquad (27)$$

Further, it can be determined that

$$z = \frac{S_{21}(\eta_s + \eta_o)}{(\eta_s + \eta_o) - S_{11}(\eta_s - \eta_o)} \qquad (28)$$

From (1), we can write

$$k_s = \frac{j}{d} \ln z \qquad (29)$$

In other words,

$$k_s = \frac{1}{d} [-\arg z + 2p\pi + j \ln |z|] \qquad (30)$$

Now the value of permeability and permittivity can be found out by using

$$\mu_s = \frac{k_s \eta_s}{\omega} \tag{31}$$

$$\varepsilon_s = \frac{k_s}{\omega \eta_s} \tag{32}$$

For resonant metamaterials, the obtained permittivity and permeability are seen to follow a Drude-Lorentz characteristic. This feature has been used extensively in the optimization of metamaterial design.

Validation of S-Parameter Retrieval Method: Using the equations given above, a Matlab code was written. In order to test the code, validation of metamaterial structures given in literature was conducted. A square split ring resonator given in (Smith et al. 2005) was designed and simulated. Let d denote the cell dimension, h be the substrate thickness, t be the copper thickness, w be the wire width, o be the outer ring length, l be the line width, g be the gap in each ring, $g1$ be the gap between the two rings. The dimensions given in this book are presented in Table 1. A dielectric substrate of $\varepsilon_r = 4.4$ and loss tangent $\delta = 0.02$ was chosen as given in this chapter.

Using these dimensions, the structure was designed as given in Fig. 9. The S-parameters were obtained and using the code for S-parameter extraction, the permittivity, permeability, wave-impedance, and refractive index were obtained and are shown in Figs. 10 and 11. The simulated results were found to match with those given in this chapter.

3.2 Design of Terahertz Metamaterials Using Scaling

Scaling is a popular technique for design of metamaterials and has been extensively reported in the literature. This is accomplished by scaling all dimensions in the metamaterial according to the wavelength of operation. However, it is noticed that the scaling operation does not always give best performance due to changes in

Table 1 Dimensions of designed metamaterial

Parameter	Value (mm)	Value in terms of wavelength (λ)
Cell dimension, d	2.5	0.0833
Substrate height, h	0.25	0.00833
Metal thickness, t	0.017	5.667×10^{-4}
Width of wire, w	0.14	0.00466
Outer ring length, o	2.2	0.0733
Line width, l	0.2	0.00667
Gap in each ring, g	0.3	0.01
Gap between rings, $g1$	0.15	0.005

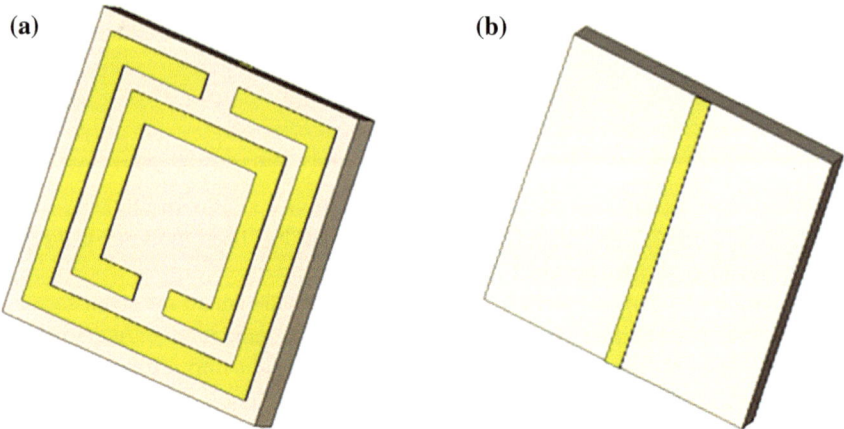

Fig. 9 Schematic of designed metamaterial. **a** Front view and **b** back view of metamaterial

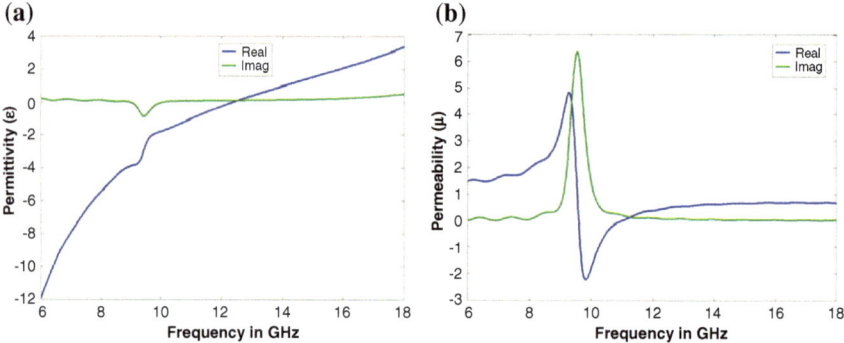

Fig. 10 Simulation results: **a** Permittivity and **b** Permeability of simulated metamaterial simulated (Fig. 9)

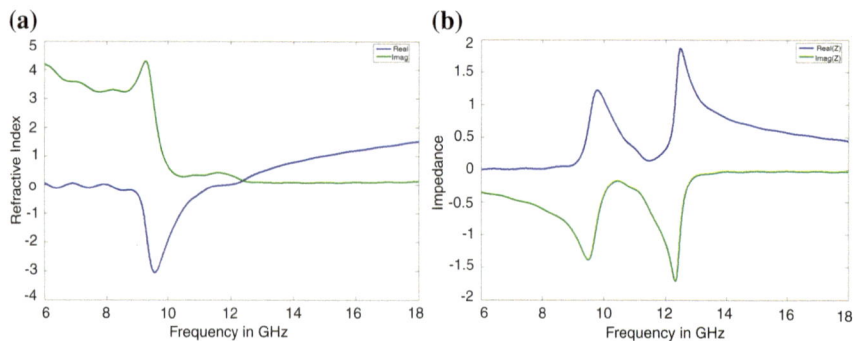

Fig. 11 Simulation results. **a** Refractive index and **b** Impedance of simulated metamaterial (Fig. 9)

Table 2 Dimensions of metamaterial obtained after scaling

Parameter	Value in terms of λ	Value for 1 THz (in μm)
Cell dimension, d	0.0833	24.99
Substrate height, h	0.00833	2.499
Metal thickness, t	5.667×10^{-4}	0.1701
Width of wire, w	0.00466	1.398
Outer ring length, o	0.0733	21.99
Line width, l	0.00667	2
Gap in each ring, g	0.01	3
Gap between rings, $g1$	0.005	1.5

material properties in the terahertz domain. The designer is able to arrive at a rough structure using this method, and optimization techniques must be used to improve the performance.

In order to validate the procedure, the design in Sect. 3.1 was scaled for operation at 1 THz, and the dimensions of which are given in Table 2. It should be noted that while the structure is scaled, the permittivity of the substrate also changes. Therefore, the value of $\varepsilon_r = 4.4$ and loss tangent $\delta = 0.02$ does not hold for 1 THz. This frequency dependence can be modeled by the Debye electric model (Beziuk et al. 2012).

$$\varepsilon_r(\omega) = \varepsilon'_\infty + \frac{\Delta\varepsilon'}{m_2 - m_1} \ln\left[\frac{\omega_2 + j\omega}{\omega_1 + j\omega}\right] \frac{1}{\ln 10} \tag{33}$$

where, ε'_∞ is the real part of permittivity at high frequency, $\Delta\varepsilon'$ is the total variation in the dielectric constant between its bounds. The lower frequency bound is denoted by ω_1 and is given as 10^{m_1}. Similarly, the upper frequency bound is denoted by ω_2 and is given as 10^{m_2}.

$$\Delta\varepsilon' = a\varepsilon_r \tan\delta(m_2 - m_1)\ln 10 \tag{34}$$

Here, ε_r and $\tan\delta$ are the dielectric constant and the loss tangent at the reference frequency respectively, ω_r (Beziuk et al. 2012).

Typically, these values are provided by the manufacturer of the dielectric in datasheets. 'a' is a constant that takes the value $\pi/2$. The high frequency permittivity is given as,

$$\varepsilon'_\infty = \varepsilon_r - a\varepsilon_r \tan\delta \ln\left(\frac{\omega_2}{\omega_r}\right) \tag{35}$$

Using these formulas, the value of permittivity at any frequency can be calculated. It has been seen that the value of permittivity decreases with frequency. The performance of the metamaterial was studied by extracting the permittivity,

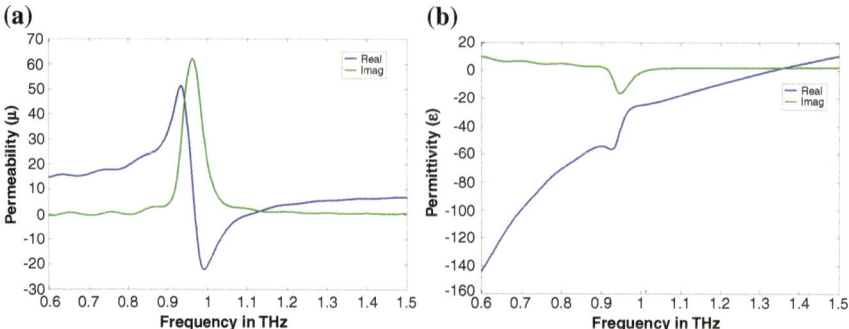

Fig. 12 Permittivity (*left*) and permeability (*right*) of scaled metamaterial

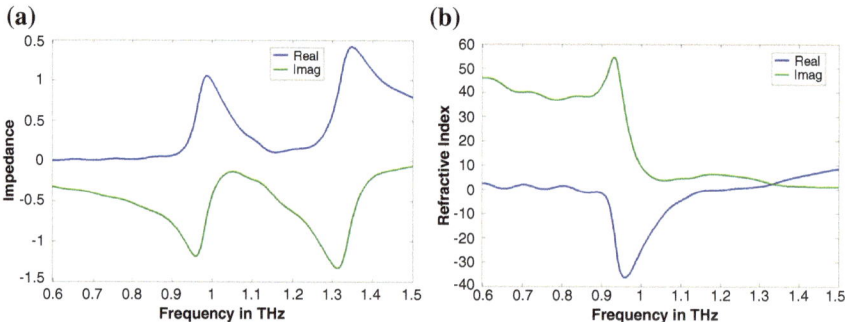

Fig. 13 Impedance (*left*) and refractive index (*right*) of scaled metamaterial

permeability, refractive index, and impedance of the metamaterial. This is shown in Figs. 12 and 13.

3.3   Absorber Design

In this section, the feasibility of scaling technique for absorber design is validated by comparing the absorption in the original metamaterial absorber and that in the scaled terahertz absorber. The absorber designs in the reference have been designed using the equations given in Sect. 2.4.

Design of Absorber for GHz Frequency: The multi-band absorber given in the reference (Shen et al. 2011) was simulated as shown. It is seen that each ring corresponds to a single absorption frequency (Fig. 14). The largest ring corresponds to the lowest resonant frequency. The square rings were made up of copper and a dielectric of relative permittivity 4 with loss tangent of 0.02 was chosen. The dimensions of the structure are given in Table 3.

Fig. 14 Multi-band
metamaterial absorber

Table 3 Dimensions of
multi-band absorber (Fig. 14)

Parameter	Value (mm)
Side of outer square (outer dimension), r_1	9.6
Side of middle square (outer dimension), r_2	7.3
Side of inner square (outer dimension), r_3	5.5
Width of each ring, w	0.5
Thickness of metal, t	0.018
Thickness of dielectric layer, h	0.78
Size of unit cell, a	10
Thickness of ground plane, g	0.018

A Matlab code was written in order to calculate Eq. (8) at different frequencies, and the absorption characteristics of the multi-band absorber was plotted as given in Fig. 15.

The designed structure resonated at three different frequencies namely 3.96, 6.68, and 9.25 GHz is due to the three different square rings, and near unity absorption was observed at these three frequencies as shown in Fig. 15. The electric field intensities at the three different frequencies were respectively plotted for each as shown in Fig. 16.

Application of Scaling Technique for Absorber: The scaling technique discussed in Sect. 3.2 was applied to the absorber designed in Sect. 3.3. However, simulation results showed that while the structure displayed resonance in the corresponding frequency band, the absorption characteristics were poor with peak absorption of about 50 %. Therefore, it can be concluded that, while scaling is an effective technique for metamaterial design, it does not produce efficient results while designing absorbers. This can be attributed to the fact that material properties of the

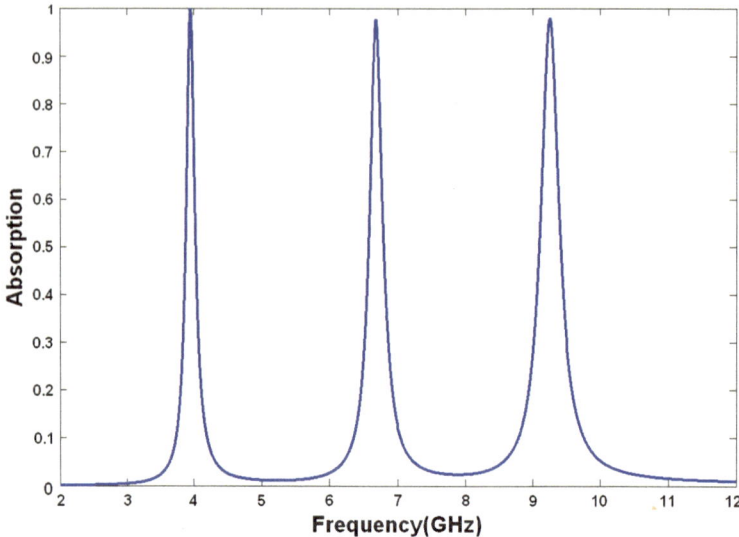

Fig. 15 Absorption in metamaterial absorber (Fig. 14)

Fig. 16 E-fields in the structure at 3.96, 6.68 and 9.25 GHz respectively

constituent elements change in the terahertz regime. Further, thickness of the substrate obtained through direct scaling may not be sufficient enough to minimize reflection.

The structure was modified by changing the thickness of the dielectric to 1.59 μm. Also, polyimide and gold were chosen as the dielectric and the metal respectively since the usage of these materials in the terahertz regime have been well documented. On simulation, three absorption frequencies, 3.48, 6.79, and 9.59 THz were observed. The absorption at these frequencies was found to be 99.5, 88.25, and 97.28 %, respectively (Fig. 17).

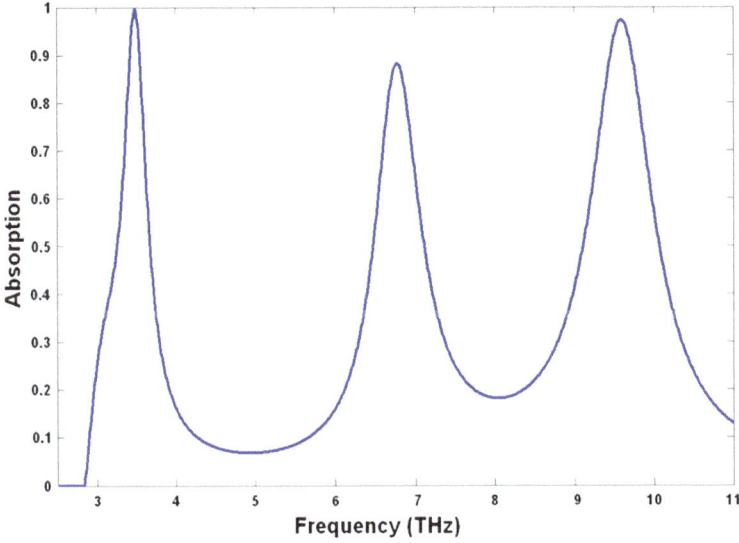

Fig. 17 Absorption characteristics of scaled absorber

3.4 Development of PSO-Based Computational Engine

From the previous sections, it is clear that the working of absorbers can be explained theoretically. However, design equations for metamaterial absorbers are not reported in the literature. Further, it is seen that addition of layers to metamaterials shifts the initial resonant frequency, a shift that cannot be predicted by equations. Therefore, EM simulations are essential to metamaterial absorber design. These simulations are carried out manually, in an iterative manner with the aim to optimize absorption at a particular frequency.

The human effort can be reduced if a computer is instructed to perform the same functions that a designer would perform manually. This feat may be accomplished using a PSO computational engine. In order to do so, the PSO algorithm is implemented on Matlab. This algorithm acts as a kernel. The calculation of the fitness is accompanied by the call of the EM tool. The PSO kernel instructs the EM tool to construct the structure, apply boundaries, simulate the structure over a particular frequency range, calculate S-parameters, and return the same to Matlab. The PSO kernel then converges to the optimal structural parameters for the desired S-parameters. This is depicted in Fig. 18.

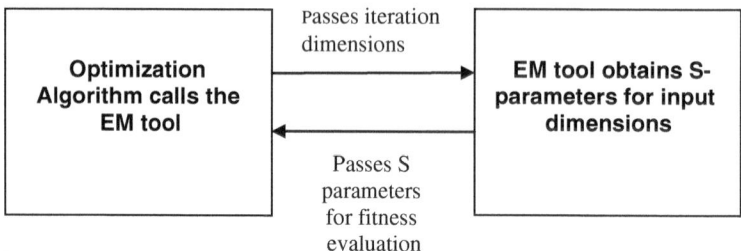

Fig. 18 Schematic of PSO-based computational engine

4 Design and Result Analysis

Using the techniques discussed in Sect. 3, a metamaterial unit cell for the absorber
for resonance at 2 THz is designed. A circular split ring resonator (CSRR) is chosen
for this design; the reasons for which are mentioned in the following sections. The
simulation results are then presented followed by the design of the absorber using
this unit cell. The absorption characterizes of the absorber are optimized by itera-
tively changing its structural dimensions. Due to the inherent narrowband nature of
resonant metamaterial-based absorbers, a technique to improve the range of operation through tuning of resonant frequency is presented.

Improvement of performance in the absorber is achieved through iterative,
manual tuning of the structural parameters. This is a very cumbersome process, and
hence it can be concluded that the above-mentioned design strategy is useful as
preliminary design strategy. In order to design high performance, accurate
absorbers, an optimization technique called PSO is then used in the form of a
computational engine that integrates a PSO kernel with a commercially available
EM solver developed specially for this purpose. The same algorithm is then used to
develop an adaptive active absorber array.

4.1 Selection of Metamaterial Structure

The CSRR resonator is a popular metamaterial structure. Its characteristics are well
known and mathematical design equations for the design of the CSRR are available.
Therefore, the CSRR at 2 THz is chosen as the metamaterial unit cell. The choice of
frequency is a result of demonstrations of skin cancer detection at that frequency by
other research groups (Pickwell and Wallace 2006).

4.2 Design of Circular Split Ring Resonator

The CSRR is then designed using the techniques given in Sect. 3.

Design of CSRR for GHz Frequency: The dimensions of the metamaterial structure, the CSRR were obtained software written by certain groups of researchers (Pradeep et al. 2011; Choudhury et al. 2013). By taking the substrate to be of lossy polyimide ($\varepsilon = 3.5$), and using the approximate dimensions given from the software, the dimensions of the CSRR were determined. For incidence along Y direction, the outer ring's external radius was found to be 5.2 mm, the width of the rings was found to be 0.3 mm, the distance between the two rings was taken as 0.3 mm, and the gap length in the two rings was determined to be 0.75 mm. Finally, the substrate dimension was found to be 12 mm × 12 mm × 0.8 mm. The structure of the SRR is given in Fig. 19 and the corresponding permeability is plotted in Fig. 20.

The choice of boundary conditions is essential to obtain accurate results in EM simulations. CSRRs show resonance when the incident wave's H-field is perpendicular to the plane of the rings. A magnetic current is generated as the rings are coupled through the gaps in the structure, resulting in negative permeability. Consequently, a perfect magnetic conductor boundary condition is assigned in the z direction. The x directions were bounded by a perfect electric conductor. The simulation shows that these boundary conditions result in Drude-Lorentz permeability characteristic curve.

Scaling of Designed SRR for 2 THz: The structure designed in Sect. 4.2 is scaled and then simulated. The transmission characteristic, S_{21} of the metamaterial CSRR structure is shown in Fig. 21a. Figure 21b is a plot of the extracted relative permeability of the metamaterial unit cell. As expected, while the scaling technique

Fig. 19 Circular split ring resonator for 2 GHz

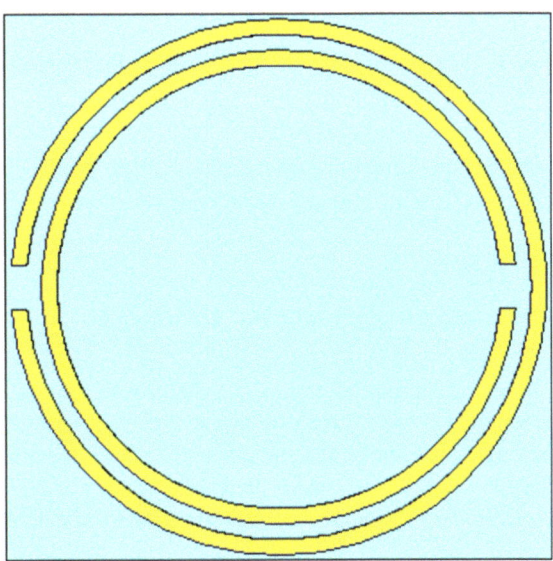

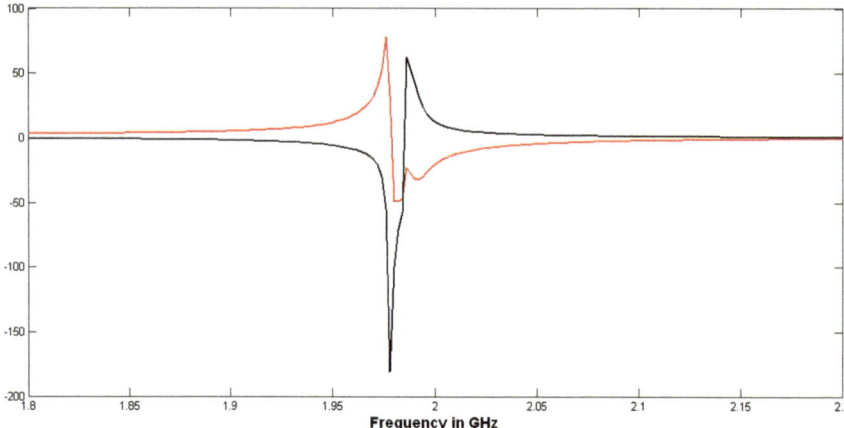

Fig. 20 Permeability of designed circular SRR for 2 GHz. The curve follows a Drude-Lorentz characteristic

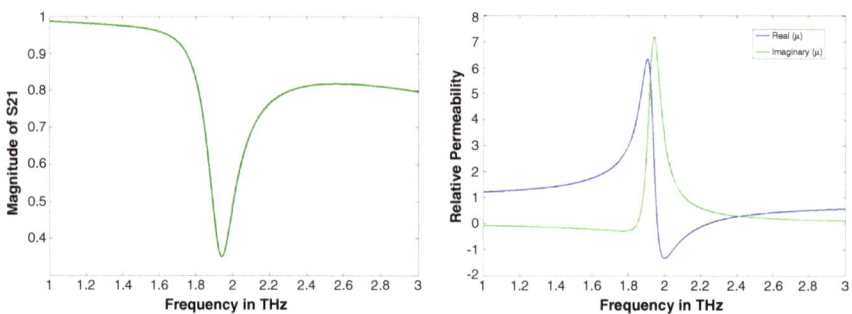

Fig. 21 Simulation results for 2 THz SRR. **a** Reflection, S21. **b** Real and imaginary parts of relative permeability

resulted in a structure that resonated at the desired frequency, the resonance characteristics were not good. Therefore, steps were taken to improve the resonance.

4.3 Design of Absorber Using SRR

The metamaterial unit cell was then used to design absorbers using the techniques given in Sect. 2.4. The absorber consists of three layers: the metamaterial unit cell, a substrate layer, and a ground plane. The dimensions of these layers must be selected properly in order to obtain good performance.

Design of Absorber for 2 THz: The metamaterial unit cell was then used to design absorbers. The absorber consists of three layers: the metamaterial unit cell, a

Fig. 22 Schematic of optimized absorber

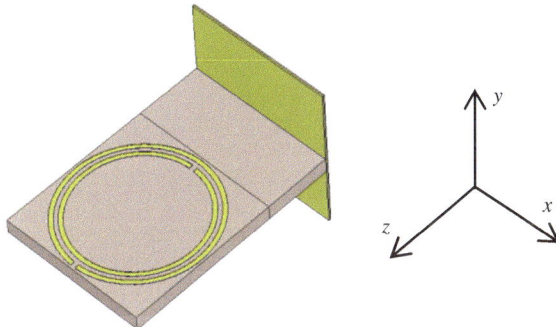

substrate layer, and a ground plane. The dimensions of these layers must be selected properly in order to obtain good performance.

The boundary conditions assigned in the simulations emulate periodic boundaries. In this case, the periodic boundaries are equal to the length and width of the substrate. It has been established that decrement in the pitch increases the absorption. Therefore, in order to improve the absorption, another polyimide dielectric was placed in between the designed metamaterial and the ground plane as shown in Fig. 22. The pitch of the unit cell was reduced to 10.7 μm in the x direction. The dielectric slab placed in between the split ring resonator and the ground plane had a length of 4.7 μm.

The best absorption was obtained when the outer ring radius was taken as 5.2 μm, the width of each ring was taken as 0.3 μm, the distance between the rings was taken as 0.275 μm, the gap length in each ring was taken as 0.35 μm, height of substrate was taken as 0.8 μm, and thickness of metal layer (for each ring) was taken as 0.01 μm. A ground plane of dimension 10.7 μm × 8.2 μm × 0.1 μm was assigned to the ring with a polyimide dielectric support placed behind it.

The absorption at 2 THz for the structure shown in Fig. 22 was found to be 98.93 % as shown in Fig. 23. On close observation, it is seen that for practical, rugged applications, the space above and below the CSRR must be filled with a material without properties similar to air such as Styrofoam.

PSO for Absorber Design: Manually determining the dimensions of the metamaterial and additional substrate layer for heighted absorption is a time-consuming and iterative task, and requires a lot of effort by the design. This human effort can be reduced by using the PSO computational engine developed in Sect. 3.4. The following parameters were chosen to be variables in the PSO algorithm: radius of outer ring, width of each ring, distance between the two rings, gap in each ring, height of substrate layer, and thickness of material in the direction of propagation. The substrate was made of polyimide and was of dimensions 10.876 × 10.876 (μm²). The thickness of the metallic rings is 0.01 μm. The algorithm is aimed to maximize absorption at 2 THz.

After optimization, the dimensions of the absorber were found to be as shown in Table 4.

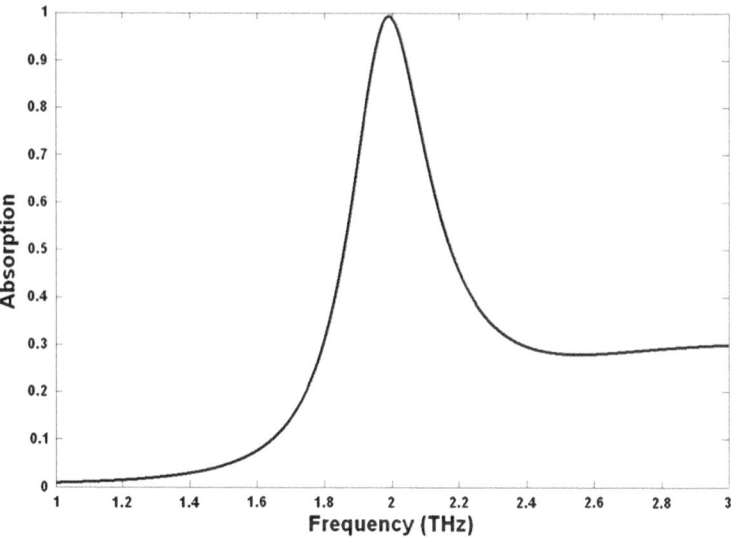

Fig. 23 Absorption characteristic curve of absorber (Fig. 22)

Table 4 Dimensions obtained after PSO optimization of 2 THz absorber

Parameter	Value (μm)
Outer radius of ring	5.2880
Width of each ring	0.3529
Gap in each ring	0.3388
Distance between the two rings	0.2907
Height of substrate	0.7900
Thickness of second substrate in direction of wave propagation	5.1976

For the PSO, 5 particles were used and the algorithm was run for 50 iterations. The total time taken was 10 h.

This fitness function was taken to be

$$f = S_{11}|_{2\,\text{THz}} \tag{36}$$

The variation of the fitness function with respect to the number of iterations during PSO is given in Fig. 24a. The resultant absorption characteristics are given in Fig. 24b. From the simulation, the absorption at 2 THz was found to be near unity.

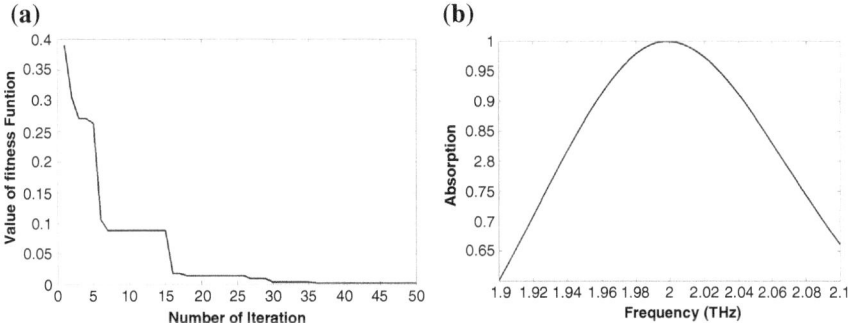

Fig. 24 PSO optimization: **a** variation of fitness function with respect to number of iterations, **b** absorption characteristics of optimized structure

4.4 Design of Active Absorber Array

The single-celled absorber developed in the previous section is used to develop a 3×2 absorber as shown in Fig. 25. However, due to the resonant nature of the SRR, a frequency tuning mechanism must be introduced in order to develop a robust, practical design.

4.4.1 Selection of Tuning Mechanism

Both thermal excitation and photoexcitation are not feasible from the view of integration with bolometric detectors as these energies might appear as noise to the detector. Electrical actuation involves changing the properties of the substrate through bias voltages. This would require specialized materials to be used as substrate such as barium strontium titanate (BST) (Bian et al. 2014), etc. The properties of such materials are inconsistent with the current design.

Hence, MEMS-based tuning is best suited for this design. The resonant frequency of a CSRR depends on the position of the gaps in the rings with respect to each other (Saha and Siddiqui 2009). The resonant frequency of resonance changes with rotation of the inner by an angle θ (Fig. 26). The new resonant frequency can be obtained using Eq. (37) by solving Eqs. (38)–(40).

$$f_o = \frac{1}{2\pi \sqrt{2 r_o L \dfrac{(\pi + k)^2 - \theta^2}{2(\pi + k)} C_{\text{pul}}}} \tag{37}$$

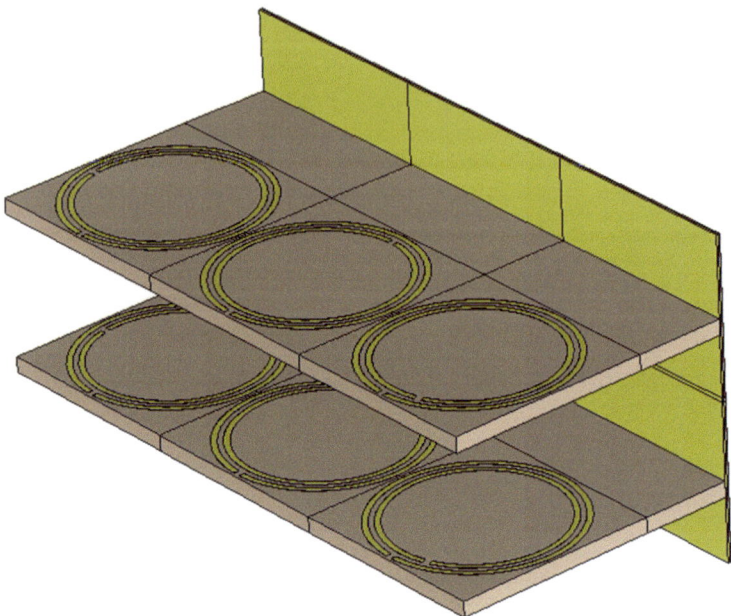

Fig. 25 3 × 2 absorber array with CSRR unit cell

Fig. 26 Rotation of inner
ring by an angle θ with
respect to outer ring

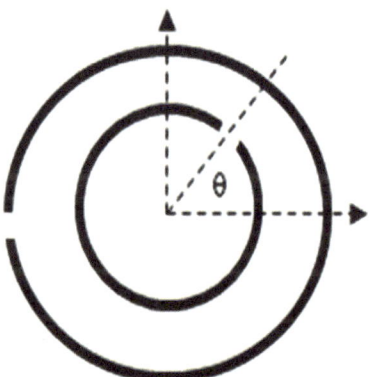

where, L is the inductance of CSRR (Eq. 40), r_o is the radius of the inner ring, C_{pul} is the per unit length capacitance between the rings, and k is a value calculated as:

$$k = \frac{C_g}{r_o C_{\text{pul}}} \tag{38}$$

where, gap capacitance, C_g is given as

$$C_g = \frac{\varepsilon_o wh}{g},\tag{39}$$

where h is substrate thickness, w is the width of the rings, and g is the length of the gap in the rings. The values for L may be calculated from

$$L = 0.00508l\left(2.303 \log_{10}\frac{4l}{d} - \gamma\right)\tag{40}$$

Here, $\gamma = 2.451$; l and d are length and width of the wire, respectively.

4.4.2 Implementation of Tuning Mechanism

The tuning mechanism can be implemented by introducing a series of MEMS switches placed at discreet angles on the outer ring of the SRR. Rotation of the outer ring may then be mimicked by switching off any one switch.

Each unit cell in the 3×2 absorber array was fitted with a switches placed at $15°$ apart. The activation of these switches in a unit cell is taken to be independent of the other unit cells. The PSO-based computational engine was then used to determine the switches that must be activated in each unit cell for resonance at 2.8 THz. Since there are four unit cells, the dimension of each particle was taken to be 4. Furthermore, the co-ordinates were allowed to take only integer values from [0, 12] in order to denote the position of the switch in between $0°$ and $180°$. Before passing these values to the EM solver, they are multiplied by 15 (as the switches are placed at an angular distance of $15°$ on the inner ring). The EM solver then creates the structure based on structural parameters obtained in the previous section, rotates each ring by the angle that is given as the input, simulates and obtains the S-parameters, and passes these S-parameter values back into the PSO kernel. The fitness is then determined by using Eq. (36).

The computational engine is run for 8 PSO particles and 20 iterations. The variation in the minimum fitness function is stored and plotted as shown in Fig. 27a, and the corresponding absorption characteristic of the optimized design in Fig. 27b. The best possible absorption characteristic obtained is 99 % at 2.8 THz.

The angles of rotation for various frequencies can be obtained using the PSO computational engine and stored as a database into the memory of a micro-controller in an active absorber array system. The absorber array can then be controlled from the processor for the resonance at the desired frequency. This concept would form the basis of the adaptive array and is depicted pictorially in Fig. 28.

Sensitivity Analysis: The output of the PSO computational engine (while determining the optimal dimensions) is highly accurate real numbers with the fractional part extending up to the fourth decimal.

While such numbers are acceptable from a simulation point of view, a study on whether such values may be achieved must be carried out.

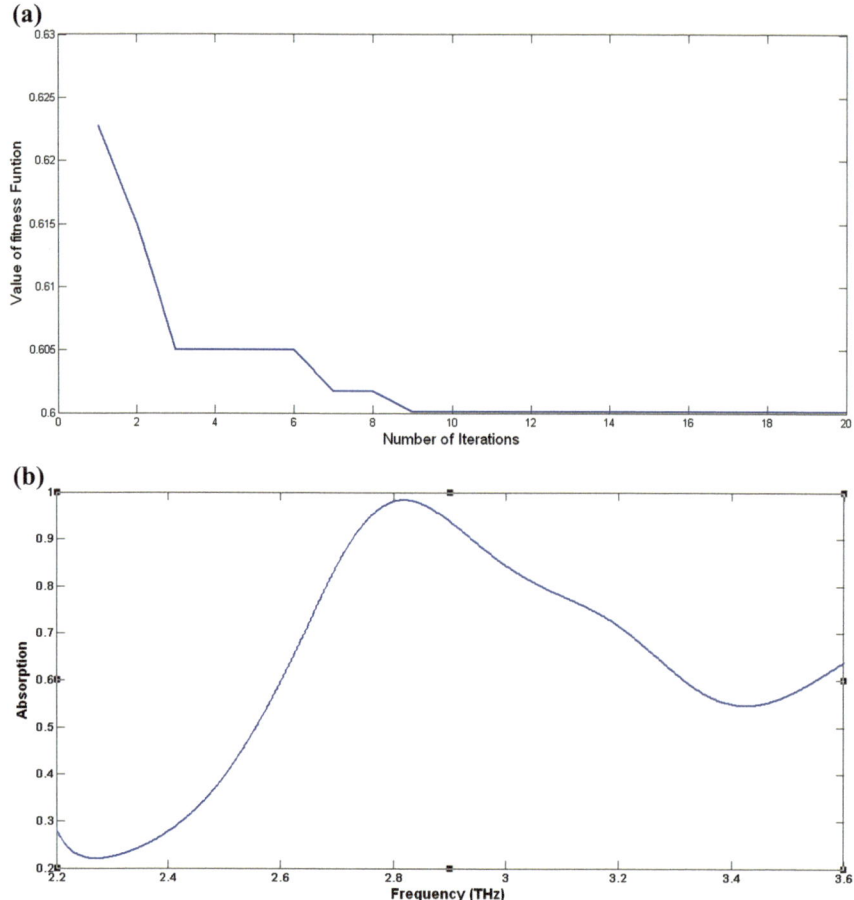

Fig. 27 a Variation of fitness function with iterations for determination of angular rotation of inner rings in 3 × 2 absorber array, **b** absorption characteristics of absorber array

Fabrication Tolerances: The most popular fabrication technique, photolithography is capable of achieving feature sizes up to 2–2.5 μm. As a result, this technique cannot be used for the fabrication of the absorber with dimensions obtained from the PSO computational engine. Therefore, it is advised to use electron beam lithography for the fabrication of the CSRR absorber. It must be noted that a 5–10 % fabrication tolerance must always be considered during fabrication.

Material Impurities: A 5 % shift in the material properties (in this case, permittivity of the substrate) must be considered in order to compensate for material impurities. Rugged applications require the usage of aluminum instead of gold for the fabrication of the CSRR as gold is susceptible to wear and tear.

The above-mentioned material and fabrication tolerances are acceptable as the absorption remains greater than 90 % in the bandwidth of operation (Fig. 29).

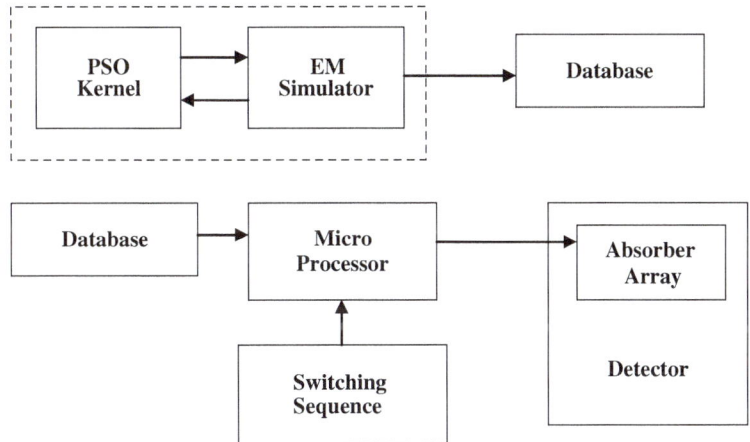

Fig. 28 Implementation of adaptive tuning for absorber array

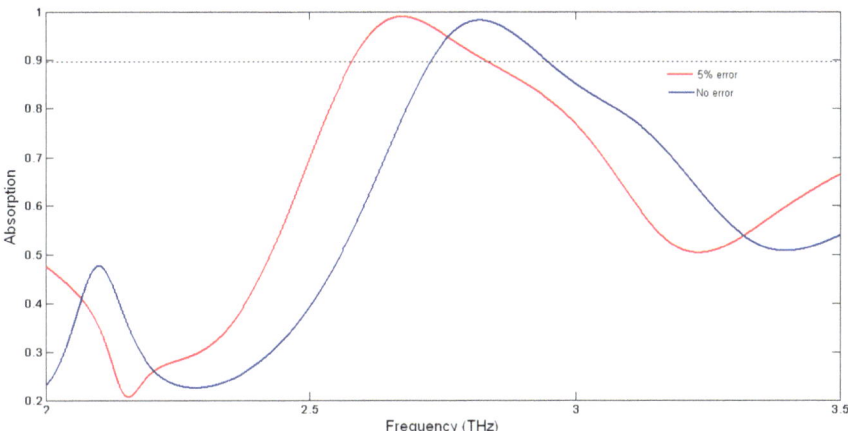

Fig. 29 Variation of absorption considering different tolerances during fabrication

5 Conclusion

Active terahertz metamaterial absorbers for biomedical applications were successfull. Absorbers with CSRR as unit cell were designed for two sample frequencies commonly used in biomedical THz-TDS—1–3 THz. The designed absorbers showed near unity absorption at the desired frequencies.

The inherent narrowband nature of the SRR-based absorbers was modified by introducing a frequency tuning mechanism in the design. The design involved rotation of the inner ring of the SRR with respect to the outer ring. Practically, this can be achieved by placing MEMS switches at equal intervals on the ring and switching them in order to emulate rotation. This mechanism enables a frequency sweep up to 0.5 THz.

Further, a PSO-based computational engine, which integrates a PSO kernel with a commercial EM solver, was developed in order to optimize the structural parameters of the absorber for high performance. This computational engine eliminates the need for manually determining the best structural parameters and is especially useful when the design equations are not well known. In addition to structural optimization, the computational engine was also used to determine the optimum inner ring rotational angles for a three element absorber array. This engine is capable of determining the rotational angles for different resonant frequencies. These values constitute a database which can be used for the rapid, adaptive tuning of the absorber array.

Further, a sensitivity analysis was also performed on the designed absorbers in order to determine the possible fabrication issue in practical situations. Although the design and simulation studies achieved the desired objectives, it is by no means the end of the research in this particular area. The following topics may be considered toward practical industrial applications as well as mass production:

Fabrication and Characterization of Structure: The optimized terahertz metamaterial absorber may be fabricated and tested in order to study the accuracy of the design procedure and the PSO-based computational engine. The characterization of the structure will give an insight into any possible deviations from the desired result. Furthermore, an attempt can be made to fit the fabricated absorber into a micro-bolometer detector and check the improvement in the performance of the detector.

Parallel Computing for Faster Results: The current version of the PSO-based computational engine was seen to take roughly 10 h to arrive at the result. The computational time may be decreased by modifying the code and executing the same in a parallel fashion.

Extension of Design Technique for Other Metamaterial Unit Cells and Materials: This project explores the design of the CSRR-based absorber. The same techniques used for the design viz. the PSO-based computational engine, may be used to design terahertz absorbers using other metamaterial geometries such as electric ring resonator (ELC), fish-net geometry, I-shaped geometry, etc. Further, different types of tuning mechanisms and substrates like BST may be considered for innovative designs.

References

Alaee, R., C. Menzel, C. Rockstuhl, and F. Lederer. 2012. Perfect absorbers on curved surfaces and their potential applications. *Optics Express* 20(16): 18370–18376.

Alves, F., B. Kearney, D. Grbovic, and G. Karunasiri. 2012. Narrowband terahertz emitters using metamaterial films. *Optics Express* 20(19): 21025–21032.

Arslanagić, S., T.V. Hansen, N.A. Mortensen, A.H. Gregersen, O. Sigmund, R.W. Ziolkowski, and O. Breinbjerg. 2013. A review of the scattering parameter extraction method with clarification of ambiguity issues in relation to metamaterial homogenization. *IEEE Antennas and Propagation Magazine* 55(2): 91–106.

Beziuk, G., P. P. Jarzab, K. Nowak, E. F. Plinski, M. J. Walczakowski, and J. S. Witkowski. 2012. Dielectric properties of the FR-4 substrates in the THz frequency range. Proceedings of 37th International Conference on Infrared, Millimeter and Terahertz Waves, Wollongong, 2pp., Sept. 2012.

Bian, Y., C. Wu, H. Li, and J. Zhai. 2014. A tunable metamaterial dependent on electric field at terahertz with barium strontium titanate thin film. Applied Physics Letters 104: 042906-1–042906-4.

Butler, L. A. 2012. Design, simulation, fabrication, and characteristics of terahertz metamaterial. Master of Science dissertation, 53pp., University of Alabama.

Caloz, C. 2006. *Electromagnetic metamaterials: transmission line theory and microwave applications*. Hoboken: Wiley. ISBN-10: 0-471-66985-7, 350pp.

Chakravarty, S., R. Mittra, and N.R. Williams. 2001. On the application of the microgenetic algorithm to the design of broad-band microwave absorbers comprising frequency-selective surfaces embedded in multilayered dielectric media. *IEEE Transactions on Microwave Theory and Techniques* 49: 1050–1059.

Choudhury, B., and R. M. Jha. 2013. PSO based Optimization Package for Metamaterial Split Ring Resonator Configurations. Registration No.: SW-7474/2013.

Choudhury, B., S. Bisoyi, and R. M. Jha. 2012. Emerging trends in soft computing for metamaterial design and optimization. Computers, Materials & Continua 31(3): 201–228 (Invited Review Paper).

Chowdhury, D. R., R. Singh, J. F. O'Hara, H. T. Chen, A. J. Taylor, and A. K. Azad. 2012. Photo-doped silicon in split ring resonator gap towards dynamically reconfigurable terahertz metamaterial. CLEO Technical Digest, 2pp.

Cui, T. J., D. R. Smith, and R. Liu. 2010. Metamaterial Theory and Design. Springer, New York, ISBN: 978-1-4419-0573-4, 367pp.

de Maagt, P. 2007. Terahertz technology for space and earth applications. International Workshop on Antenna Technology: Small and Smart Antenna, Metamaterials and Applications, pp. 111–115.

Ekmekci, E., K. Topalli, T. Akin, and G. T. Sayan. 2009. A tunable multi-band metamaterial design using micro-split SRR structures. Optics Express 17(18): 16046–16058.

Ferguson, B., and X.C. Zhang. 2009. Materials for terahertz science and technology. *Nature Materials* 1: 26–33.

Goudos, S.K., and J.N. Sahalos. 2006. Microwave absorber optimal design using multi-objective particle swarm optimization. *Microwave and Optical Technology Letters* 48(8): 1553–1558.

He, S. 2013. Broadband THz absorbers with graphene-based anisotropic metamaterial films. IEEE Transactions on Terahertz Science and Technology 7pp.

Hokmabadi, M.P., D.S. Wilbert, P. Kung, and S.M. Kim. 2013. Terahertz metamaterial absorbers. *Terahertz Science and Technology* 6(1): 40–58.

Huang, W., W. Lin, L. Wang, L. Chang, and C. Liao. 2012. Broadband optimization for a terahertz metamaterial absorber. Microwave Workshop Series on Millimeter Wave Wireless Technology and Applications (IMWS), pp. 1–3, Sept. 2012.

Humphreys, K., J. P. Loughran, M. Gradziel, W. Lanigan, T. Ward, J. A. Murphy, C. O'Sullivan. 2004. Medical applications of Terahertz Imaging: a Review of Current Technology and Potential Applications in Biomedical Engineering. Proceedings of the IEEE International Conference on Engineering in Medicine and Biology Society, pp. 1302–1305.

Jin, N., and Y. R. –Samii. 2007. Advances in particle swarm optimization for antenna designs: real-number, binary, single-objective and multiobjective implementations. *IEEE Transactions on Antennas and Propagation* 55(3): 556–567.

Kearney, B. T. 2013. Enhancing microbolometer performance at terahertz frequencies with metamaterial absorbers. Doctorate of Philosophy dissertation, 69pp. Naval Postgraduate School 2013.

Kennedy, J., and R. Eberhart. 1995. Particle swarm optimization. Proceedings of IEEE International Conference on Neural Networks, pp. 1942–1948.

Kowerdziej, R., M. Olifierczuka, B. Salskib, and J. Parkaa. 2012. Tunable negative index metamaterial employing in-plane switching mode at terahertz frequencies. *Liquid Crystals* 39 (7): 827–831.

Kozlov, D. S., M. A. Odit, I. B. Vendik, Y.-G. Roh, S. Cheon, C. Lee, W. 2011. Tunable terahertz metamaterial based on resonant dielectric inclusions with disturbed Mie resonance. *Applied Physics A: Material Science and Processing*, pp. 465–470.

Kung, P., and S. M. Kim. 2013. Terahertz metamaterial absorbers for sensing and imaging. Proceedings of Progress in Electromagnetics Research Symposium, pp. 232–235, Taipei.

Li, B., L.X. He, Y.Z. Yin, W.Y. Guo, and X.W. Sun. 2012. A symmetrical dual-band terahertz meta-material with cruciform and square loops. *Progress in Electromagnetics Research C* 33: 259–267.

Moser, H. O., B. D. F. Casse, O. Wilhelmi, and B. T. Saw. 2005. Terahertz response of a microfabricated rod–split-ring-resonator electromagnetic metamaterial. *Physical Review Letters* 063901-1–063901-4.

Němec, H., P. Kužel, F. Kadlec, C. Kadlec, R. Yahiaoui, and P. Mounaix. 2009. Tunable terahertz metamaterials with negative permeability. *Physical Review Letters* 79: 241108-1–241108-4.

Ozbey, B., and O. Aktas. 2011. Continuously tunable terahertz metamaterial employing magnetically actuated cantilevers. *Optics Express* 19(7): 5741–5752.

Padilla, W. J., A. J. Taylor, C. Highstrete, M. Lee, and R. D. Averitt. 2006. Dynamical electric metamaterial response at terahertz frequencies. Proceedings of the 15th International Conference, Pacific Grove, USA, pp. 642–644, Aug 2006.

Parsopoulos, K.E., and M.N. Vrahatis. 2002. Recent approaches to global optimization problems through particle swarm optimization. *Natural Computing* 1: 235–306.

Paul, O., C. Imhof, B. Lägel, S. Wolff, J. Heinrich, S. Höfling, A. Forchel, R. Zengerle, R. Beigang, and M. Rahm. 2009. Electrically tunable metamaterial for polarization-independent terahertz modulation. Conference on Lasers and Electro-optics, pp. 1–2, June 2009.

Pickwell, E., and V. P. Wallace. 2006. Biomedical Application of terahertz technology. *Journal of Physics D: Applied Physics*, 39(17).

Pradeep, A., S. Mridula, and P. Mohanan. 2011. Design of an edge–coupled dual-ring split-ring resonator. *IEEE Antennas and Propagation Magazine* 53(4).

Pratibha, R., K. Park, I.I. Smalyukh, and W. Park. 2009. Tunable optical metamaterial based on liquid crystal-gold nanosphere composite. *Optics Express* 17(22): 19459–19469.

Saha, C., and J. Y. Siddiqui. 2009. Estimation of the resonance frequency of the conventional and rotational circular split ring resonators. Proceedings of Applied Electromagnetics Conference (AEMC), pp. 1–3, Dec. 2009.

Shadrivov, I.V., S.K. Morrison, and Y.S. Kivshar. 2006. Tunable split-ring resonators for nonlinear negative-index metamaterials. *Optics Express* 14(20): 9344–9349.

Shen, X., T.J. Cui, J. Zhao, H.F. Ma, W.X. Jiang, and H. Li. 2011. Polarization-independent wide-angle triple-band metamaterial absorber. *Optics Express* 19(10): 9401–9407.

Siegel, P.H. 2004. Terahertz technology in biology and medicine. *IEEE Transactions on Microwave Theory and Techniques* 52(2): 2438–2447.

Smith, D.R., J.B. Pendry, and M.C.K. Wiltshire. 2004. Metamaterials and negative refractive index. *Science* 305: 788–792.

Smith, D. R., D. C. Vier, T. Koschny, and C. M. Soukoulis. 2005. Electromagnetic parameter retrieval from inhomogeneous metamaterials. *Physical Review*, 71: pp. 036617-1–036617-11.

Tao, H., A. Strikwerda, C. Bingham, W. J. Padilla, X. Zhang, and R. Averitt, D. 2008. Dynamical control of terahertz metamaterial resonance response using bimaterial cantilevers. In: PIERS Proceedings, pp. 870–873.

Vendik, I.B., O.G. Vendik, M.A. Odit, D.V. Kholodnyak, S.P. Zubko, M.F. Sitnikova, P.A. Turalchuk, K.N. Zemlyakov, I.V. Munina, D.S. Kozlov, V.M. Turgaliev, A.B. Ustinov, Y. Park, J. Kihm, and C.-W. Lee. 2012. Tunable metamaterial structures for controlling THz radiation. *IEEE Transactions on Terahertz Science and Technology* 2(5): 538–549.

Vinoy, K. J., and R. M. Jha. 1996. Radar Absorbing Materials from Theory to Design and Characterization. Kluwer Academic Publishers, Boston. ISBN:0-7923-9753-3.

Yang, T., X. Li, and W. Zhu. 2013a. A tunable metamaterial absorber employing MEMS actuators in THz regime. NEMS, pp. 829–832, China, Apr. 2013.

Yang, T., X. Li, and W. Zhu. 2013b. A tunable metamaterial absorber employing MEMS Actuators in THz regime. 8th IEEE International Conference on Nano/Micro Engineered and Molecular Systems (NEMS), pp. 829–832.

Zadeh, L. A. 1992. Fuzzy logic, neural networks and soft computing. One-page course announcement of CS 294-4, Spring 1993. The University of California at Berkeley, Nov. 1992.

Zhang, W., W. M. Zhu, H. Cai, P. Kropelnicki, A. B. Randles, M. Tang, H. Tanoto, Q. Y. Wu, J. H. Teng, X. H. Zhang, D. L. Kwong, and A. Q. Liu. 2013. A tunable MEMS THz waveplate based on isotropicity dependent metamaterial. *Transducers* 538–541.

Zhu, W. M., A. Q. Liu, T. Bourouina, D. P. Tsai, J. H. Teng, X. H. Zhang, G. Q. Lo, D. L. Kwong, and N. I. Zheludev. 2012. Micro-electromechanical Maltese-cross metamaterial with tunable terahertz anisotropy. *Nature Communications*, 6pp. Dec. 2012.

About the Book

This book describes a metamaterial-based active absorber for potential biomedical engineering applications. Terahertz (THz) spectroscopy is an important tool for imaging in the field of biomedical engineering, due to the non-invasive, non-ionizing nature of terahertz radiation coupled with its propagation characteristics in water, which allows the operator to obtain high-contrast images of skin cancers, burns, etc. without detrimental effects. In order to tap this huge potential, it is important to build highly efficient biomedical imaging systems by introducing terahertz absorbers into biomedical detectors. The biggest challenge faced in the fulfilment of this objective is the lack of naturally occurring dielectrics, which is overcome with the use of artificially engineered resonant materials, viz. metamaterials. This book describes such a metamaterial-based active absorber. The design has been optimized using particle swarm optimization (PSO), eventually resulting in an ultra-thin active terahertz absorber. The absorber shows near unity absorption for a tuning range of terahertz (THz) application.

© The Author(s) 2016   43
B. Choudhury et al., *Active Terahertz Metamaterial for Biomedical Applications*,
SpringerBriefs in Computational Electromagnetics,
DOI 10.1007/978-981-287-793-2

Author Index

A
Akin, T., 12
Aktas, O., 12
Alaee, R., 11
Alves, F., 11
Arslanagic, S., 18–20
Averitt, R., 12
Averitt, R.D., 12
Azad, A.K., 12

B
Beigang, R., 12
Beziuk, G., 23
Bian, Y., 33
Bingham, C., 12
Bisoyi, S., 4, 14
Bourouina, T., 12
Breinbjerg, O., 18–20
Butler, L.A., 13

C
Cai, H., 2, 12
Caloz, C., 7
Casse, B.D.F., 10
Chakravarty, S., 15
Chang, L., 11
Chen, H.T., 12
Cheon, S., 12
Chowdhury, D.R., 12
Cui, T.J., 9

D
de Maagt, P., 2

E
Eberhart, R., 15
Ekmekci, E., 12

F
Ferguson, B., 2
Forchel, A., 12

G
Goudos, S.K., 15
Gradziel, M., 5
Grbovic, D., 11
Gregersen, A.H., 18–20
Guo, W.Y., 11

H
Hansen, T.V., 18–20
He, L.X., 11
He, S., 8
Heinrich, J., 12
Highstrete, C., 12
Höfling, S., 12
Hokmabadi, M.P., 9
Huang, W., 11
Humphreys, K., 5

I
Imhof, C., 12

J
Jarzab, P.P., 23
Jiang, W.X., 24
Jin, N., 15, 16

K
Kadlec, C., 12
Kadlec, F., 12
Karunasiri, G., 11
Kearney, B.T., 2, 5, 11
Kennedy, J., 15
Kholodnyak, D.V., 12

Kihm, J., 12
Kim, S.M., 9
Kivshar, Y.S., 12
Koschny, T., 21
Kowerdziej, R., 12
Kozlov, D.S., 12
Kropelnicki, P., 12
Kung, P., 9
Kužel, P., 12
Kwong, D.L., 12

L
Lägel, B., 12
Lanigan, W., 5
Lederer, F., 11
Lee, C.-W., 12
Lee, M., 12
Li, B., 11
Li, H., 24
Li, X., 8, 12
Liao, C., 11
Lin, W., 11
Liu, A.Q., 12
Liu, R., 9
Lo, G.Q., 12
Loughran, J.P., 5

M
Ma, H.F., 24
Menzel, C., 11
Mittra, R., 15
Mohanan, P., 29
Morrison, S.K., 12
Mortensen, N.A., 18–20
Moser, H.O., 10
Mounaix, P., 13
Mridula, S., 29
Munina, I.V., 12
Murphy, J.A., 5

N
Němec, H., 13
Nowak, K., 23

O
Odit, M.A., 12
O'Hara, J.F., 12
Olifierczuka, M., 12
O'Sullivan, C., 5
Ozbey, B., 12

P
Padilla, W.J., 12
Park, K., 12

Park, W., 12
Park, Y., 12
Parkaa, J., 12
Parsopoulos, K.E., 16, 20
Paul, O., 12
Pendry, J.B., 2
Pickwell, E., 28
Plinski, E.F., 23
Pradeep, A., 29
Pratibha, R., 12

R
Rahm, M., 12
Randles, A.B., 12
Rockstuhl, C., 11
Roh, Y.-G., 12

S
Saha, C., 33
Sahalos, J.N., 15
Salskib, B., 12
Samii, Y.R., 15, 16
Saw, B.T., 10
Sayan, G.T., 12
Shadrivov, I.V., 12
Shen, X., 24
Siddiqui, J.Y., 33
Siegel, P H., 2
Sigmund, O., 18–20
Singh, R., 12
Sitnikova, M.F, 12
Smalyukh, I. I., 12
Smith, D.R., 2, 9, 21
Soukoulis, C.M., 21
Strikwerda, A., 12
Sun, X.W., 11

T
Tang, M., 12
Tanoto, H., 12
Tao, H., 12
Taylor, A.J., 12
Teng, J.H., 12
Topalli, K., 12
Tsai, D.P., 12
Turalchuk, P.A., 12
Turgaliev, V.M., 12

U
Ustinov, A.B., 12

V
Vendik, I.B., 12
Vendik, O.G., 12

Vier, D.C., 21
Vinoy, K.J., 13
Vrahatis, M.N., 16, 20

W

Walczakowski, M.J., 23
Wallace, V.P., 28
Wang, L., 11
Ward, T., 5
Wilbert, D.S., 9
Wilhelmi, O., 10
Williams, N.R., 15
Wiltshire, M.C.K., 2
Witkowski, J.S., 23
Wolff, S., 12
Wu, C., 33
Wu, Q.Y., 12

Y

Yahiaoui, R., 13
Yang, T., 9, 12
Yin, Y.Z., 11

Z

Zadeh, L.A., 14
Zemlyakov, K.N., 12
Zengerle, R., 12
Zhai, J., 33
Zhang, W., 12
Zhang, X., 12
Zhang, X.C., 2
Zhang, X.H., 12
Zhao, J., 24
Zheludev, N.I., 12
Zhu, W., 9, 12
Ziolkowski, R.W., 18, 20
Zubko, S.P., 12

Subject Index

A
Absorber design, 24
Active absorber array, 33
Adaptive tuning, 4

B
Biomedical imaging, 3

C
Circular split ring resonator, 29

M
Metamaterial absorbers, 13
Metamaterials, 6
    group velocity, 8
    left-handed materials, 7
    phase velocity, 8
    transmission line equivalent, 7

N
Non-resonant metamaterials, 9

P
Particle swarm optimization, 14

fitness, 15
global best, 15
particle, 15
personal best, 15

R
Resonant metamaterials, 9

S
S-parameter retrieval, 18
    permeability, 18
    permittivity, 18

T
Terahertz instruments, 5
Terahertz radiation, 2
Terahertz region, 1
Terahertz time domain spectroscopy, 5
Tuning mechanism, 11, 33
    electrical actuation, 12
    photoexcitation actuation, 12
    thermal actuation, 12

© The Author(s) 2016
B. Choudhury et al., *Active Terahertz Metamaterial for Biomedical Applications*,
SpringerBriefs in Computational Electromagnetics,
DOI 10.1007/978-981-287-793-2